Springer Series in Physical Environment

12

Günter Groß

Numerical Simulation of Canopy Flows

With 127 Figures

Springer-Verlag
Berlin Heidelberg New York
London Paris Tokyo
Hong Kong Barcelona
Budapest

Professor Dr. Günter Groß
Institut für Meteorologie und Klimatologie
Universität Hannover
Herrenhäuser Str. 2
W-3000 Hannover 1
Germany

ISSN 0937-3047
ISBN-13:978-3-642-75678-8   e-ISBN-13:978-3-642-75676-4
DOI: 10.1007/978-3-642-75676-4

Library of Congress Cataloging-in-Publication Data. Groß, Günter, 1954– Numerical simulation of canopy flows / Günter Groß.   p.   cm. — (Springer series in physical environment : 12) Includes bibliographical references (p.   ) and index. ISBN-13:978-3-642-75678-8
1. Forest meteorology — Mathematical models.   2. Plant canopies — Mathematical   models.
3. Deforestation — Environmental aspects — Mathematical models.   4. Air Flow —
Mathematical models.   5. Mesometeorology — Mathematical models.   I. Title.   II. Series.
SD390.5.G76 1993 551.69152 — dc20 93-2799

© Springer-Verlag Berlin Heidelberg 1993
Softcover reprint of the hardcover 1st edition   1990

The use of general descriptive names, registered names, trademarks, etc. in this publication does not imply, even in the absence of a specific statement, that such names are exempt from the relevant protective laws and regulations and therefore free for general use.

Typesetting: Macmillan India Ltd., Bangalore-25
32/3145/SPS-5 4 3 2 1 0 – Printed on acid-free paper

Preface

After scanning through scientific journals and popular-scientific magazines published in recent years, it becomes evident that forests have moved into the focus of interest. This may be ascribed to a considerable extent to increasing forest damage illustrating even to the occasional hiker the dangers to which the eco-system forest are exposed.

The existence of forest trees has always been threatened by natural factors such as bacteria and fungi, drought and frost. This has led to lethal damage to a complete forest only in exceptional cases. However, this situation has changed dramatically in recent years and, as a result of anthropogenic influences, the very existence of forests is endangered.

One of the contributing causes is the growing pollution of the atmosphere with a host of noxious substances. Despite the lack of final, unequivocal proof, it is safe to state that these noxious, anthropogenic compounds play a prominent role in the origin of the forest damage observed. A forest, which at present appears to be completely intact, may die off within only a few years.

Clearance of a wooded area may convert the underlying ground within much shorter periods into fallow land. Although the climatic changes expected as a result of the deforestation of large areas, predominantly in tropical rain forests, are of a more long-term nature, such changes may materialize at the actual site of deforestation from one day to the next. Especially in the densely populated areas of central Europe, large areas are cleared of their forest cover in order to make way for additional residential and industrial buildings.

Such a dramatic change in land utilization will lead to a marked change in the local climate. It is necessary to assess these influences in a qualitative and quantitative manner prior to the actual deforestation, in order to minimize the potential adverse effects by appropriate precautions.

In the present book, a new method of estimating such effects is introduced. Various scenarios may be envisaged for the deforestation of smaller areas without a single tree being actually being cut down with the aid of numerical simulation. The evaluation of the results thus obtained may allow important conclusions as to the most likely changes to be expected in local climate. Local parameters such as topography may be taken into account in the process. This represents a notable advantage of this new method over the other methods applied so far.

The model may also be used to control damage already caused by changes in land utilization. An example is the optimal arrangement of windbreak hedges and tree rows designed to reduce wind velocities and thereby soil erosion.

Other uses not explicitly referred to in this book will emerge during practical application of the methods outlined. Other problems might become solvable only after some modification or expansion of the model. Consequently, the model introduced here should be considered only as a very basic tool for the solution of problems such as those encountered in meso-scale meteorology, town and regional planning, forestry and agriculture.

I am especially indebted to Prof. F. Wippermann for the numerous fruitful discussions on the subject and for his constructive criticism of the manuscript. Thanks are also due to Springer-Verlag for including this book in their prestigious publishing program.

Hannover, March 1993   Günter Groß

Contents

1 Introduction

In temperate climatic regimes, trees and forests represent an important aspect of the natural environment. Forests serve as areas of recreation for the population suffering the noise and air pollution in urban cities. Silence, fresh air and a climate without extremes are some of the factors found in forests, but which are notably lacking in cities.

The recreational nature of a forest is a feature of this ecosystem most easily appreciated by each individual. There are, however, also a multitude of other functions fulfilled by forest, such as protection against erosion, filtering of water and air, noise and wind protection, leveling of climatic extremes, and a source of raw materials. This illustrates that questions regarding our very existence are closely linked to the stability of forest systems.

In temperate latitudes, the forest as a whole represents the stable end-state of an ecosystem under given conditions of climate and soil. This will be established when production and reduction processes in the forest are in balance. Larger deviations from this equilibrium, caused by severe frost or periods of drought, are usually compensated fairly rapidly. However, when the average environmental conditions are changed, a destabilization of the forest ecosystem will ensue.

Even a stable end-state is subject to various perils resulting from natural and anthropogenic causes. In addition to the natural hazards to the existence of a forest, caused by fires, animals, insects, bacteria and lichens, the influence of human civilization is also increasing. Initially, this was from the regular cultivation of agricultural plants which demanded increasing areas of land. To satisfy this demand, frequently wide forest areas had to be cleared. The establishment of settlements and the construction of larger infrastructural features such as water reservoirs, industries, airports, and roads led to the disappearance of even more forests. Only very recently, the deforestation of areas closer to cities has been considered in planning concepts, e.g., in Freiburg at the foot of the Black Forest. In this case an improved supply of fresh air from the surrounding region into the city itself is expected.

An additional adverse factor is the increase in air pollution since the start of industrialization in the last century. Whereas, initially, the resulting damage was restricted to forests in the immediate vicinity of the industrial centers, more recently, severe damage has also been observed in regions farther from the source of the pollutants. Tall stacks emit pollutants into the atmosphere and these are then transported over hundreds of kilometers, becoming chemically modified on the way. To the best of our present knowledge, these air pollutants represent an important contributing factor to the complex phenomena of forest decline (Papke et al. 1986).

Due to the interaction of natural and anthropogenic factors large forest areas may disappear, leading to climatic changes from a local to a global extent. The tropical rainforests, which cover about 10% of the land surface of the earth, and account for some 40% of the global forest cover, are of prime importance for the global climate. The deforestation of larger areas will immediately result in increased soil erosion, probably leading to dramatic changes in the water budget of the soils. This, in turn, may have devastating effects on the ecosystem of the Tropics which is adapted to the present conditions. A reduction in the areas covered by rainforests may, in turn, reduce evapotranspiration, thus notably disturbing the hydrological cycle in the troposphere. Compared with the effects on a local scale, these global changes take place comparatively slowly. In both cases, however, potential quantitative and qualitative changes in climate have to be considered.

In order to investigate the effects of changing land use on, e.g., local climate, field experiments alone are inadequate, since two situations must be known, i.e., the situation prior to and the situation after the disappearance of the trees. The use of numerical simulation models is an important tool for investigating potential climatic changes. Deforestation of a wide area, hard to reverse in reality, may be easily carried out in the model. By comparing the results of the simulations with and without forest, realistic assessments of the most likely changes in climatic patterns are possible.

On a global scale, this has to be investigated by use of rather complex models which take into account the general atmospheric circulation with good spatial resolution and which also considers the interaction of the subsystems of ocean, inland ice and land surfaces. The unraveling of this interaction is still fraught with great difficulties, as the typical time constants of the subsystems are of completely different orders of magnitude, and many constraining parameters are not well known or understood.

The climatic changes are strongly influenced on a regional scale by human activities. Such regional and local changes are frequently more readily understood by taking field measurements.

In the following chapters a numerical model is described which permits an estimation of the consequences of deforestation on local climatic conditions by considering the specific site parameters. The potential of the model and its limitations will be illustrated for a number of examples.

Before describing the actual model, observations concerning the variation in meteorological variables in and above stands of trees will be discussed. This will facilitate the evaluation of the applicability of the simulation results in comparison with the observed data.

A forest is made up of individual trees, each of which influences the wind pattern close to the ground. Therefore, the flow pattern around an isolated tree will be investigated first. Effects of changes in stand geometry will then be investigated together with those effects which may be ascribed to variations in meteorological parameters.

In a further step towards a model forest, changes in flow patterns resulting from groups of individual trees will be discussed, permitting the characterization of a complete stand by certain parameters. This, in turn, facilitates the inclusion in the model of the influence of tall stands on the distribution of meteorological variables without considering individual trees in detail.

This approach will then be used to investigate air flows in and above taller stands. Furthermore, the numerical model will be expanded to include the influence exerted by topographic features together with diurnal variations in wind, temperature and humidity.

Extensive, fairly homogeneous forests in a flat area may be studied by one-dimensional models. In this case a variation in meteorological variables is only allowed in the vertical direction. These calculations provide valuable indications as to the sensitivity of the results to changes in stand-specific parameters such as tree height, type of trees and stand density.

These numerical experiments, however, do not allow the study of spatial and temporal changes in meteorological patterns which result from the change in land utilization from one type to another, e.g., from forest to open ground. Therefore, in order to investigate realistic areas (Fig. 1.1), it is necessary to carry out three-dimensional simulations. Two examples of this type of approach will be described.

On the occasion of the extension of Frankfurt International Airport by the addition of a runway in the early 1980s, deforestation of a large area became necessary. Potential effects of this change in land use on the local climate were estimated by the Deutscher Wetterdienst (DWD: German weather service). This expertise was based on field observations gathered prior to the deforestation together with data on the differences found between the distribution of meteorological variables over forest and open ground. However, the complex interactions between changed land utilization and the resulting new wind patterns can be estimated only very roughly, if at all. Mesoscale simulation models, like the one introduced here, significantly contribute to the solution of these problems.

The results of the respective simulations illustrate the marked changes in the meteorological parameters over the deforested area. These changes are not surprising as the forest was replaced by a concrete runway and surrounding meadows. The most important findings are the pronounced increase in wind speed throughout the day and an increase in the diurnal temperature variation. With increasing distance from the clearance these changes become attenuated, and from about 400 m onwards they are no longer discernible. This leads to the conclusion that the deforestation carried out for the runway strongly modifies the local climate, but the influence of only this change in land use on a regional scale is negligible.

A situation becomes multifaceted and more complex than in the above example when the area of interest is not flat but is orographically structured. The structure and intensity of thermally induced slope wind patterns (e.g.

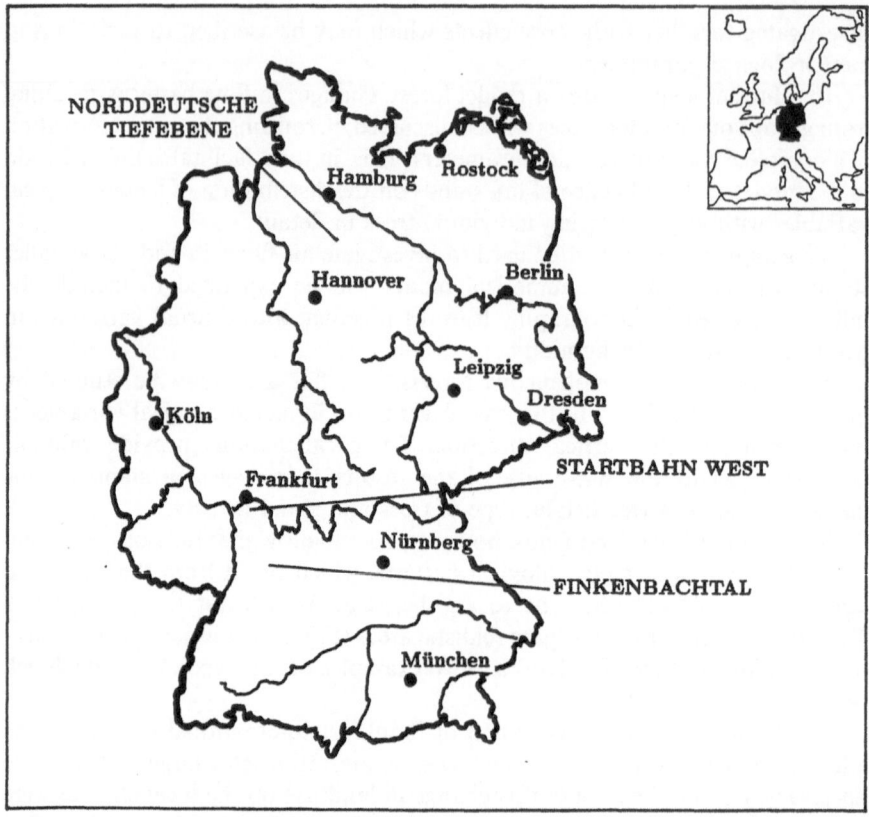

Fig. 1.1. Sketch map of Federal Republic of Germany showing the location of the study areas

nocturnal downslope winds) will be changed completely. This problem is treated in a second example in which possible climatic changes resulting from a simulated deforestation of the Finkenbach valley in the southwestern part of Germany were investigated. This particular site was selected as meteorological data provided by the DWD during a field experiment were available for comparison.

A particular aspect of this example is the role played by forested slopes on the nocturnal production of cold air. It was found that bare slopes produce a much smaller volume of cold air than slopes do covered by tall stands. This is of great importance in urban and regional planning as the urban climate is strongly influenced by the supply of fresh air from the surrounding areas.

2 Distribution of Meteorological Variables in Forest Canopies

2.1 Forest Structure and Utilization

The biosphere represents the interface between the atmosphere and lithosphere on land. It is, as it were, the zone within which the various forms of life occur. It extends vertically for a few dozen meters into the atmosphere (the tallest trees measuring about 100 m in height) and for a few meters into the ground, which is inhabited by animals and microorganisms.

Plants are of special importance in the biosphere as they account for about 99% of all living matter on earth. In the following we shall focus on the largest of the higher plants, namely trees.

An individual tree can be subdivided into three parts: crown, trunk and roots. The crown is the upper part of the tree with branches and twigs and offers optimal exposure of the leaves and needles to the environment. In this respect the arrangement of the leaves for the optimal reception of radiation is of prime importance, as it is radiation which represents the source of energy for a plant and which is thus the prime factor controlling the plant's development. However, radiation sometimes also represents a stress factor. One will therefore encounter the densest concentration of needles or leaves where the radiation conditions permit the plant to make optimal use of the radiation.

An objective parameter describing the density of the foliage is the *leaf area density*, b, i.e., the leaf area per unit volume of dimensions $m^2 m^{-3}$. Due to the three-dimensional structure of an individual tree, this parameter is a function of space. However, usually only the vertical profile of b is determined. The vertical integral of the leaf area density is the *leaf area index*, L, which gives the ratio between the total area of all leaves and the horizontal projection of the tree section.

In deciduous trees often with flat-topped crowns, the majority of the leaf mass is found in the uppermost part of the crown, with the amount decreasing rapidly downwards (Fig. 2.1a). In the more cone-shaped crowns of conifers, the maximum of b is found much lower in the crown (Fig. 2.1b). In Table 2.1 the leaf area index of various types of vegetation is given. For forests, the data are relatively scarce due to the problems inherent to the determination of L (Halldin 1985). This also explains the fairly wide range of individual values for L. In addition, the seasonal variations of L, which has a maximum value when foliage is fully developed in the summer and a minimum value just prior to the appearance of new leaves, have also to be taken into account.

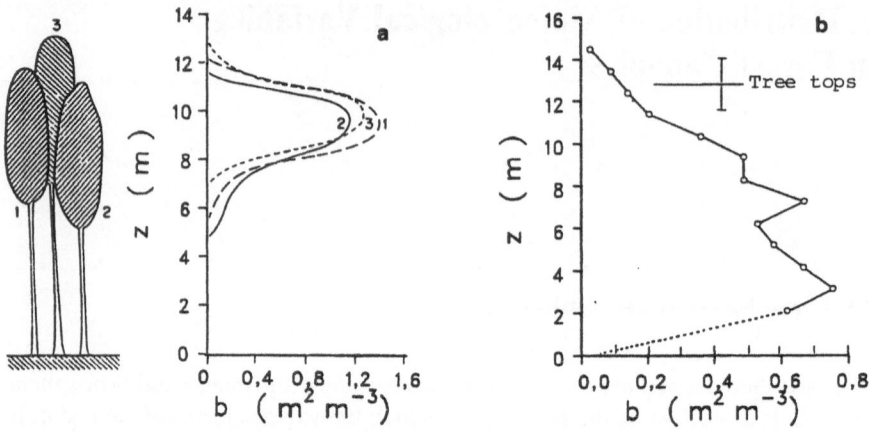

Fig. 2.1. Typical vertical profiles of leaf area density. **a** Deciduous tree (after Rauner 1976); **b** conifer. (After Hicks et al. 1975)

Table 2.1. Leaf area index for different types of vegetation. (After Whittaker and Likens 1975)

Type of vegetation	Leaf area index L ($m^2\,m^{-2}$)	
	Mode	Range
Tropical rainforest	8	6–16
Summer-green forest	5	3–12
Evergreen temperate forest	12	5–14
Boreal forest	12	7–15
Dry bush	4	4–12
Savannah	4	1–5
Agricultural plants	4	4–12

The trunk of a tree represents the connection between the crown and the ground, and determines the overall shape of the tree by its specific structure. It permits the upward transport of water and nutrients, necessary for producing organic compounds (assimilates) from the roots. Assimilates are the basic material for building roots, leaves, wood and bark.

For the forester the trunk represents the economically most important portion of the tree. The subterranean root system supplies the plant with water and the nutrients dissolved in it. In particular, the uptake of water from the soil facilitates a balanced water budget for a plant, without which a controlled advance of the various processes of life would be impossible. The above-ground parts of a plant have to transpire continuously in order to avoid overheating in the midday' sun. Evaporation increases with the vapor pressure gradient between the evaporating surface and ambient air. However, even in water-saturated air there is still some transport of water through the plant, as the essential transport of nutrients must not be interrupted.

Different individual trees growing in close vicinity to one another result in a general type of vegetation referred to as forest. While, even nowadays, in the Tropics and in Subarctic climates dense, almost impenetrable rainforests and coniferous forests develop, consisting of a multitude of different types of vegetation, such primeval, unadultered forests in central Europe are only preserved in poorly accessible parts of mountain ranges.

The majority of stands here are plantations introduced by man, consisting frequently of only one species of tree. Although such carefully managed forest utilization may be advantageous, the dangers inherent to this approach are best illustrated by the increased damage to such monocultures by wind or pests.

Such damage is not only of direct economic importance, but may also affect the surrounding areas. Thus, a forest inhibits soil erosion because it is a good wind shelter and leads to a pronounced reduction of wind close to the ground and also because of the close interlacing of soil and roots. Furthermore, trees provide important protection against avalanches.

The regulating influence of forests on the circulation of water is of even greater importance. Surface runoff is greatly reduced by the tremendous water storage capacity of forest soils. The moisture taken up by the soil is partly returned to the air during dry days. In this way significant amounts of water

Table 2.2. Water budget of different vegetation stands. (After Larcher 1984)

Type of stand	Location	Precipitation (mm/year)	Evapotranspiration (% of precipitation)	Runoff (surface and ground water in % of precipitation)
Tropical rainforest	Zaire	1900	73	27
Savannah	Zaire	1250	82	18
Deciduous forest	Europe	600	67	33
	Asia	700	72	28
Coniferous forest	Europe	730	60	40
	Eastern Europe	800	65	35
Mountain forest	Alps	1640	52	48
	Europe	1000	43	57
	Andes	2000	25	75
	America	1300	38	62
Savannah	Tropics	700	85	15
Grassland	Europe	700	62	38
Steppe	Eastern Europe	500	95	5
Semi-desert	Subtropics	200	95	5
Tundra	America	180	55	45

vapor are recycled as shown in Table 2.2. Suitable afforestation programs thus facilitate optimal control of the water budget of a certain area. If, for example, in drier areas the largest possible portion of the precipitation is to be retained for groundwater replenishment, afforestation patterns should be rather widely spaced so that evaporation is reduced and the soil is protected against erosion, at the same time the soil should be rather porous so that water can easily penetrate.

Due to the great number of leaves with a large surface area, a tree represents an excellent filter for aerosols and gases. The total area formed by the leaves and branches of a tree with a crown diameter of 6 m and a leaf area index of 10 is about 300 m^2. The deposition of aerosols and dust is controlled by the wind velocity V and the sink velocity. As V is fairly low within the stand, notable deposition may take place on the leaves. This will be the case especially in the crowns and along the margins of stands. In contrast, gases are taken up directly into the cells during CO_2 exchange by the plant. Due to the filtering action of a forest the air in its interior is less polluted than the air over the open ground surrounding it.

This filtering effect, so advantageous to man, may pose a grave danger for the plant itself. The gases taken into the cells can lead to damage which greatly impairs the proper functioning of the leaves. The noxious substances deposited on the tree will be washed into the soil, where they will eventually be taken up through the root system into the nutrient cycle of the tree. These factors can lead to a weakening of the resistance of a tree and even to its death. The dramatic effects of these processes are illustrated by observations on air pollution and forest damage (Schöpfer and Hradetzky 1984).

In addition to their protective and regulatory properties, forests also exert a strong influence on local climates. Due to the slowdown caused by the large number of roughness elements, the wind speed within a stand, up to the tree height, is much lower than over the surrounding open ground. Diurnal wind variations are also much lower within the stand. As a result of low albedo, forests absorb a large portion of the direct solar radiation reaching them. However, since the main part of this radiation is not absorbed in a plane (as over bare soil) but in a layer of several meters, ranging from the ground up to the tree top, one can observe a reduced warming of the atmosphere above the stand. At the same time the air within the stand during the day may be up to 10 °C cooler than the air outside. Finally, because of evaporation the relative humidity inside the stand is higher than over the surrounding open fields and has smaller amplitudes. Forest climates are thus characterized by a reduction in the extremes of various climatic parameters.

Despite the pronounced influence of forest on local climates, very little is known about the extent of global interrelations. In particular, the vast deforestation presently being carried out in the tropical rainforests represents a dramatic impact on the global CO_2 budget, the effects of which are difficult to quantify (Baumgartner and Kirchner 1980).

2.2 Variation of Meteorological Parameters Within a Stand

2.2.1 Radiation

On its passage through the earth's atmosphere, solar radiation is subjected to a number of modifications before it eventually reaches the top of the stand in an attenuated form. In addition to this direct radiation, diffuse irradiance and longwave downward radiation of the atmosphere also contribute to the radiation flux from above the stand; together these represent the radiative input. The output consists of reflected shortwave solar radiation and the longwave emission of the forest itself. The sum of radiative input and output is referred to as the radiation budget, R_N. During the daytime it is mostly positive, and during the night it is mostly negative.

The absorption of radiation over open ground takes place close to the earth's surface in a fairly thin layer. In a forest, a fairly thick zone frequently takes part in these processes, with the upper third of the canopy being of particular importance. Naturally, this only applies to a fairly dense crown cover as otherwise part of the radiation can penetrate to the ground, leading to a transformation of radiation also at this level.

The radiation penetrating the stand is successively reduced by the foliage and only a fraction of the quantity measured at the top of the canopy will eventually reach the ground. The degree of reduction within the stand depends on the intensity of the radiation directly above the forest (R_N) and on the vertical distribution of the stand elements, which is characterized by $b(z)$ and $L(z)$.

As determined by Baumgartner (1956) the radiation budget at different levels within a thinned stand of spruces 5–6 m high is presented in Fig. 2.2. The

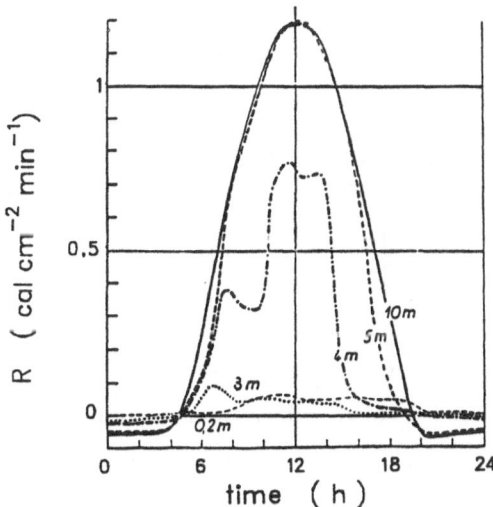

Fig. 2.2. Diurnal variation of radiation budget at five different heights in a thinned-out spruce stand. (After Baumgartner 1956)

diurnal variation of R_N above the forest (10 m) and in the treetop zone (5 m) shows little difference. In the crown zone (4 m) there is already a notable reduction which becomes more pronounced as lower levels are measured. During the day, at 20 cm above the ground less than 10% of the original value of R_N at 10 m is recorded.

Mitscherlich et al. (1967) presented measurements of the vertical distribution of the relative illumination intensity in different stands. The term 'relative' in this context relates to the same parameter in open ground. In a stand composed of oaks and beeches there is a rather regular decrease in the illumination intensity almost down to the ground (Fig. 2.3a). The horizontal bars represent the range of the individual measurements. Their rather uniform width illustrates the uniform distribution of the leaf mass. The same measurements for a pine forest are shown in Fig. 2.3b. In the upper half of the canopy the illumination intensity falls drastically. Below this it remains rather constant due to the lack of needles in this zone. The surface expression of a pine forest is nowhere near as uniform as that of a deciduous stand. The pointed cone shape of the pine crowns leaves large open spaces between the individual trees and a closed canopy is established only about 5–6 m below the tops. Therefore, the illumination intensity in the top zone is highly variable: open-ground values are observed in the interstices between the tops and reduced values occur in the shade of the tops themselves. This is clearly illustrated by the width of the bars in Fig. 2.3b.

The curves in Fig. 2.3 suggest that the vertical change in net radiation, R_{NP} (radiation budget), within the stand may be approximated by an exponential function according to Bouger–Lambert's law. The approach by Uchijiama (1961):

$$R_{NP}(z) = R_N\, e^{-k_c L(z)}, \qquad\qquad (2.1)$$

shows fairly good agreement with actual measurements. In this case R_N is the net

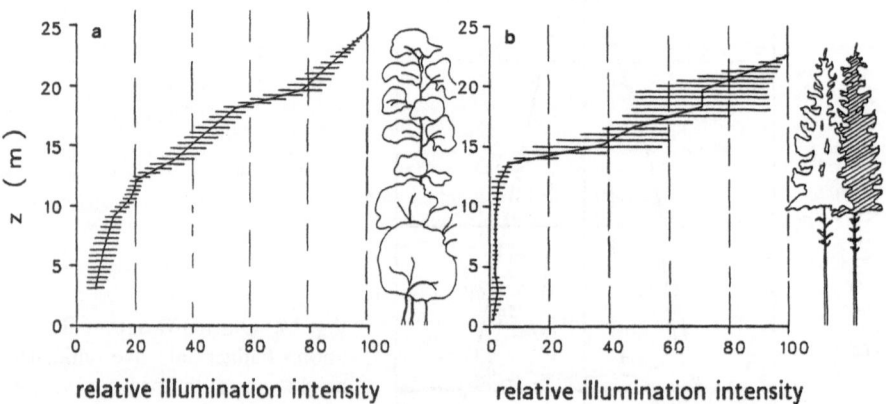

Fig. 2.3. Vertical profiles of relative illumination intensity (after Mitscherlich et al. 1967). a Deciduous forest; b pine forest

radiation directly above the stand and is virtually identical to the value over open ground. The extinction coefficient k_c is an empirical constant which shows some relation to the time of day. Landsberg et al. (1973) reported a minimum of $k_c = 0.5$ at midday, increasing to 2.5–3 in the evening. Jarvis et al. (1975) compiled further observed values of k_c. Moreover, Budagovsky et al. (1968) take the solar height into account in:

$$R_{NP}(z) = R_N e^{-G_g L(z)/\sin Z},$$ (2.2)

with an empirical constant G_g and zenith angle Z.

Impens and Lemeur (1969) report that the agreement between calculated and measured values is improved when using

$$R_{NP}(z) = R_N e^{-k_{c1} L(z) + k_{c2} L(z)^2},$$ (2.3)

with $k_{c1} = 0.622$ and $k_{c2} = 0.055$.

2.2.2 Temperature

The temperature change in the atmosphere within a stand is determined by the divergence of the energy fluxes. According to Thom (1975), the energy budget in a canopy volume may be written in a symbolic form as:

$$R_N = Q_B + Q_V + Q_H + Q_M + Q_P + Q_D,$$ (2.4)

where R_N is net radiation flux, Q_B is soil heat flux, Q_H is turbulent sensible heat flux, Q_V is turbulent latent heat flux, Q_M is energy transfer during metabolic processes, Q_P is heat storage in the phytomass and Q_D is horizontal flux of sensible and latent heat. Typical ranges of the various components of the energy budget and their signs (positive downward, negative upward) are summarized in Table 2.3.

The data show that over very long periods of time the energy budget is mainly controlled by R_N, Q_H and Q_V. A good approximation for Eq. (2.4) is thus frequently obtained by considering only these three fluxes. Only at the edge of stands may Q_D represent a notable contribution. Detailed measurements of the various energy fluxes have been recorded by, e.g., McCaughey (1985) in a 19-m-high mixed forest.

Table 2.3. Typical energy budget over a stand (W m^{-2}). (After Thom 1975)

Time	R_N	Q_B	$Q_H + Q_V$	Q_M	Q_P	Q_D
Sunrise	0	−5	−8	3	10	±5
Midday	500	25	461	12	2	±25
Sunset	0	5	3	2	−10	±15
Midnight	−50	−25	−20	−3	−2	±10

In a closed stand the energetically most important zone of transfer is situated at crown height. In this zone the largest divergence in energy fluxes, and thus pronounced changes in temperature, are encountered. On a summer day one can observe that the highest daytime temperatures are encountered in the crown, whereas at ground level they are much lower (e.g., Chroust 1968). Prior to the appearance of the leaves in spring, solar radiation penetrates almost unimpeded to the forest floor and transformation of radiation then takes place at the earth's surface. It is warmed considerably and the vertical temperature profile is similar to that developed over open ground (Fig. 2.4).

The diurnal temperature variation in a thinned spruce stand is shown in Fig. 2.5 (after Baumgartner 1956). The 5- to 6-m-high stand shows a closed canopy, and highest temperatures are developed in the upper parts and not at ground level. During the daytime, there is a strong temperature increase at the top of the stand. At about 1500 LST, the temperature difference between ground

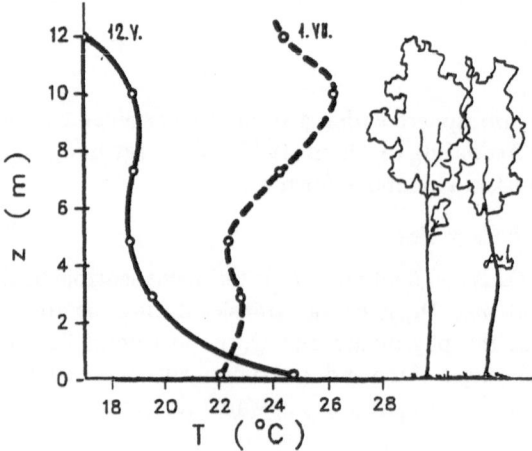

Fig. 2.4. Vertical temperature profile in an oak stand. *Solid line* Prior to; *dashed line* after appearance of the leaves. (After Chroust 1968)

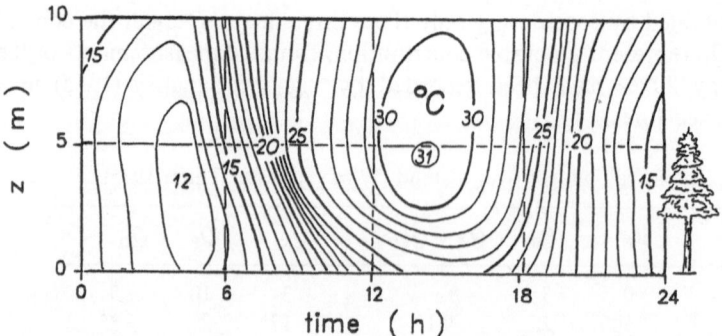

Fig. 2.5. Diurnal temperature variation in a thinned-out spruce stand. (After Baumgartner 1956)

Table 2.4. Annual mean temperature for a forest and an open field. (After Lützke 1961)

Height (m)		T_{forest} (°C)	T_{field} (°C)
23.0	Above canopy	7.36	
17.0	Middle of crown	7.18	
15.0			7.52
2.0	Trunk zone	6.70	7.36
0.25		6.59	7.32
0.002	Ground surface	6.56	7.70

level and canopy is 6°C. During the night there is virtually no difference in temperature with height, except for a weakly pronounced minimum at 0500 LST in the canopy. Similar diurnal temperature variations have been observed in pine and deciduous stands of different densities (Chroust 1968; Mitscherlich 1971). In a very open stand the diurnal variation of T represents a combination of that in a forest and that over open ground. Göhre and Lützke (1956) observed during the daytime a weak maximum at treetop level and a temperature difference of 1°C between crown and ground level, and at night a pronounced temperature inversion.

In Table 2.4 the mean annual temperatures close to the ground, over an open field and within an open pine forest are compared. It is found that at all levels measured, the air temperature was higher over the open field than in the forest. This clearly shows that the forest floor on the whole receives less heat than the open meadow.

2.2.3 Humidity

Through evaporation the stand elements in contact with the atmosphere continuously lose water. This transpiration has to be maintained since it represents the main prerequisite for the transport of nutrients within the plant. The necessary water is taken up through the root system, which extracts moisture from the soil. The water content of the soil is thus a critical supply factor for a tree.

Through the water conduits in the trunk and branches, moisture is eventually passed to the leaves. Water vapor is then lost to the air through the stomata, small openings in the leaves. This physical process of transpiration is a diffusion process which increases in intensity with the vapor pressure difference between the leaf surface and ambient air. It decreases with a higher diffusive resistance. The process is controlled by the transfer resistance at phase boundaries, the diffusion interval and the diffusion area. The turbulent energy exchange of individual leaves (index l) may, according to Denmead (1984), thus be expressed as:

$$Q_{\mathrm{HI}} = \frac{c_p \varrho (T_{\mathrm{l}} - T_{\mathrm{a}})}{r_{\mathrm{g}}}, \tag{2.5}$$

$$Q_{\mathrm{VI}} = \frac{c_p \varrho}{\Gamma} \frac{e_{\mathrm{s}}(T_{\mathrm{l}}) - e_{\mathrm{a}}}{r_{\mathrm{g}} + r_{\mathrm{s}}}. \tag{2.6}$$

Here, Γ denotes the psychrometer constant, T_{l} the leaf temperature, e_{s} the saturation vapor pressure at the leaf surface, T_{a} and e_{a} temperature and vapor pressure, respectively, of the ambient air, r_{g} the boundary layer resistance, and r_{s} bulk stomatal resistance. Neglecting the minor terms in Eq. (2.4), the energy budget of a leaf may be written as:

$$R_{\mathrm{NI}} = Q_{\mathrm{HI}} + Q_{\mathrm{VI}}, \tag{2.7}$$

with R_{NI} representing the net radiation absorbed by the leaf. Combining Eqs. (2.5)–(2.7) leads to

$$Q_{\mathrm{VI}} = \frac{\Delta R_{\mathrm{NI}} + c_p [e_{\mathrm{s}}(T_{\mathrm{a}}) - e_{\mathrm{a}}]/r_{\mathrm{g}}}{\Delta + \Gamma (1 + r_{\mathrm{s}}/r_{\mathrm{g}})}, \tag{2.8}$$

with Δ being the inclination of the saturation vapor pressure (hPa/°C) against the temperature curve $[e_{\mathrm{s}}(T_{\mathrm{l}}) = e_{\mathrm{s}}(T_{\mathrm{a}}) + \Delta (T_{\mathrm{l}} - T_{\mathrm{a}})]$.

Equation (2.8) was introduced by Monteith (1965) as an expansion of the relation given by Penman (1984). It is thus referred to as the Penman Monteith equation.

As a measure of the water vapor content of the air, the vapor pressure or the relative humidity is usually used. The diurnal variation of the vapor pressure on a clear day is represented by the wellknown double-peaked curve shown in Fig. 2.6a. The first minimum coincides with the lowest temperature, as at this time the air cannot absorb much moisture. At the same time the rate of evapotranspiration of the leaves is also low and only the soil is available as a source of moisture. After sunrise the opening stomata permit the escape of

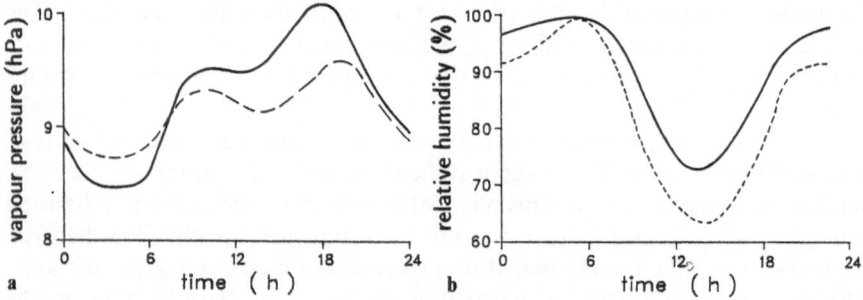

Fig. 2.6. Diurnal variation of **a** vapor pressure and **b** relative humidity 2 m above ground in a forest (*solid line*) and over a meadow (*dashed line*). (After Lützke 1967)

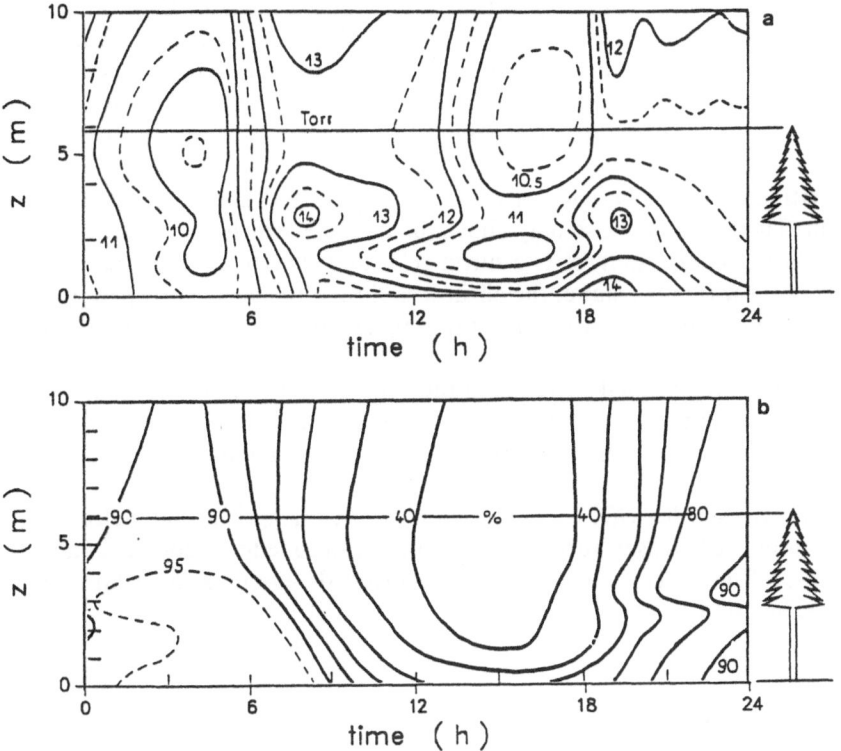

Fig. 2.7. Diurnal variation of **a** water vapor pressure and **b** relative humidity in a thinned-out spruce stand. (After Baumgartner 1956)

water vapor, leading to the pronounced rise in vapor pressure. Around midday, turbulent diffusion attains its maximum value and the vapor pressure drops due to strong mixing with dry air. A renewed rise occurs only in the afternoon when the strong turbulent diffusion dies down.

The relative humidity is mainly controlled by the temperature and consequently there is no double-peaked curve. Periods of higher temperature are correlated with low relative humidities and vice versa (Fig. 2.6b).

Investigations by Baumgartner (1956) present an impression of the diurnal variation of vapor pressure and relative humidity as a function of height (Fig. 2.7). The temporal variation in both parameters is highest at crown level, whereas in the trunk zone the amplitude of the diurnal variation is lowest.

2.2.4 Wind

The wind velocity over open, unforested ground under thermally neutral conditions shows an almost logarithmic increase with height in the lower few decameters of the atmosphere. This vertical wind profile is modified in the forest due to a pronounced retardation effect. In zones with a high concentration of stand elements, expressed by b and L, this retardation effect is strongest. With high wind velocities over open ground and a fairly loose stand, the air flow can penetrate deep into the forest, leading at times to a secondary velocity maximum.

Within a crown the wind speed is rather low and almost independent of height. There is an increase only above the tree tops which is as pronounced as over open ground. As there is a direct relation between leaf area density and wind speed within the stand, the air flow is slowed to a smaller degree prior to the appearance of the leaves than during those times when the trees are in full foliage (Fig. 2.8).

There is little diurnal variation in the wind velocity within the stand due to the lower variation in thermal stratification and as a result of the sheltering effect of the canopy, which keeps the large-scale winds from penetrating the stand.

This wind-shelter effect of higher stands is used in agricultural meteorology. By planting hedges as windbreaks soil erosion can be reduced and yields may be increased. The structure of the obstacle controls the extent of the calm zone on its leeward side.

2.3 Methods of Investigation

The distribution of the meteorological variables in time and space can be investigated by a variety of methods. The classical method is field measure-

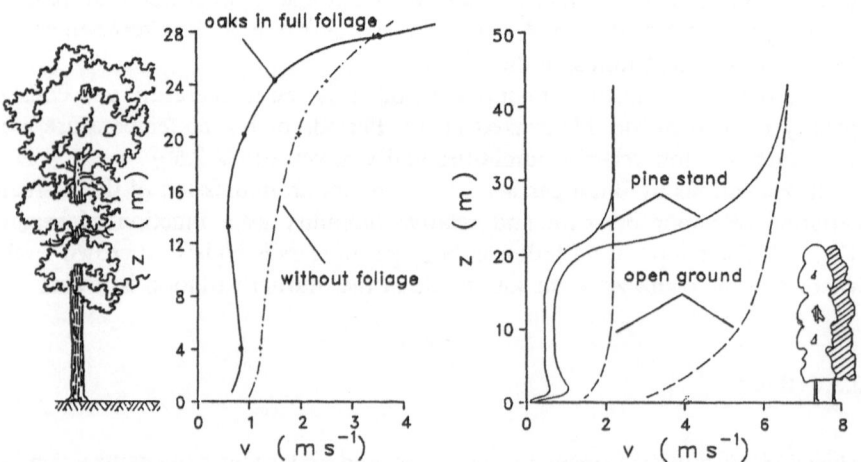

Fig. 2.8. Vertical wind speed profiles. *Left* After Fons (1940): *right* after Geiger (1961)

ments. These are increasingly accompanied by wind tunnel experiments and numerical model calculations. A selection of the results obtained with the various methods will be presented in this section.

2.3.1 Field Measurements

Data collected during field measurements should be representative of the site. Homogeneity with regard to topography and the plant stand is thus a necessary condition. This is especially the case for the windward zone of the site as the air flow above the stand should have attained a certain equilibrium. Optimum ranges for the upwind fetch and the arrangements of instruments have been compiled by Fritschen (1985) and Sceicz (1975).

Due to the large variability in tree structure, such a homogeneous area is usually difficult to find. Fritschen et al. (1969) observed considerable changes in natural stands even over relatively short distances (Fig. 2.9).

In order to obtain a complete set of data, a multitude of parameters of the system, made up of soil, stand and atmosphere, would have to be recorded. As this usually requires an excessive amount of work, most field observations are

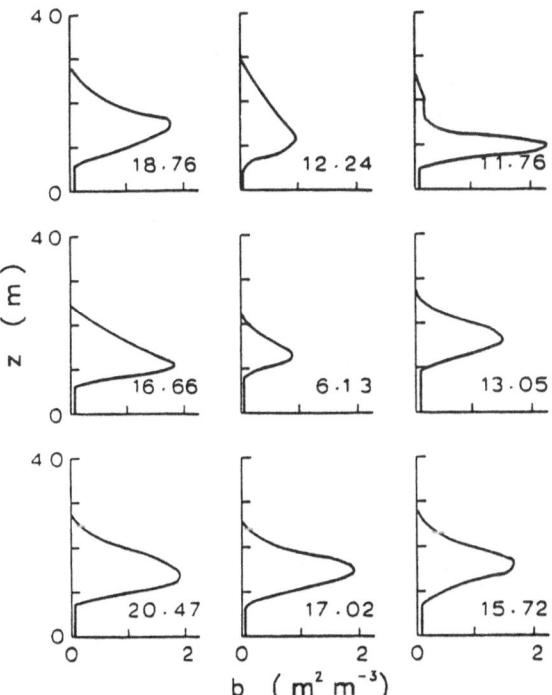

Fig. 2.9. Vertical leaf area density profiles at different locations within a stand. The leaf area index is given by the *number* in each diagram. (After Fritschen et al. 1969)

restricted to recording only those parameters which are required for solving a certain problem.

In most investigations, mean variables such as wind, temperature, humidity and radiation are directly recorded, whereas turbulence parameters are derived from them.

In Fig. 2.10 measurements of the wind velocity in various types of stands and under different conditions of thermal stratification are shown. Normalized to the wind speed recorded at crown height (which is different in all measurements), the profiles show a rather similar structure. In some of the data sets a secondary maximum close to the ground is developed, while the wind velocity remains rather constant in the canopy above.

The conditions shown do not apply to stand margins and areas of deforestation, as here the transition from open ground into the stand takes place, and this is accompanied by drastic changes. Wind blowing from open ground into a stand is slowed down. Fritschen et al. (1969) found that about 60 m inside the stand, equivalent to about two to three times the tree height, the wind has established a new equilibrium (Fig. 2.11). The wind velocity close to the ground in front of, within, and behind a 600-m-long and 28-m-high pine stand already shows a reduction of V at some distance from the forest edge. Within the first few rows of trees, the air flow in the trunk zone is accelerated, declining almost exponentially thereafter. In the lee of the forest, the original, undisturbed situation has not yet been reestablished, even at a distance equivalent to 20 times the tree height (Fig. 2.12). Observations by Flemming (1964) confirm that the wind in the marginal zones of a forest may be described by an exponential function.

Windbreaks are usually an extreme example of the above-mentioned situation when the forest is reduced to a single line of trees or is replaced by a small

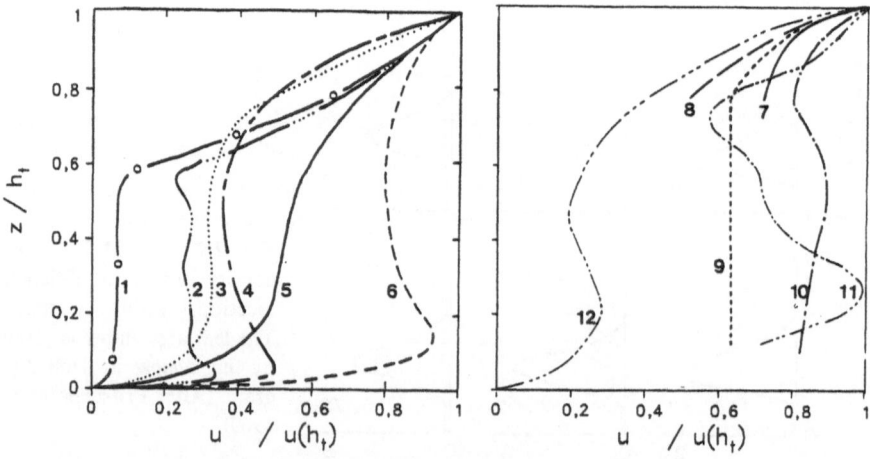

Fig. 2.10. Observed vertical profiles of normalized wind speed. (After Kurata 1982)

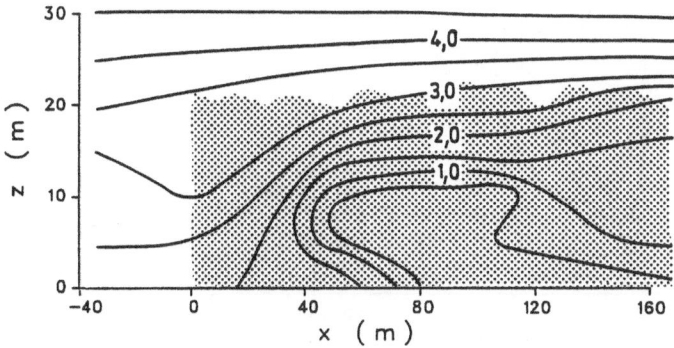

Fig. 2.11. Vertical cross section of wind speed (in m s^{-1}) near the edge of the forest. (After Fritschen et al. 1969)

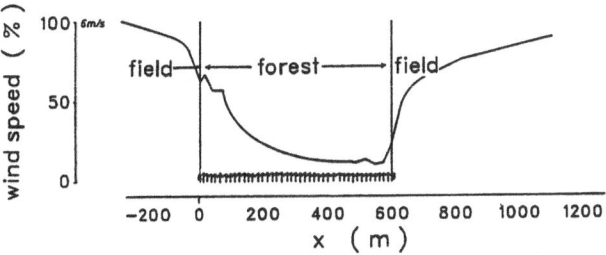

Fig. 2.12. Normalized wind speed close to the ground. (After Nägeli 1954)

artificial obstacle. In view of the importance of windbreaks in agriculture, a number of experimental investigations have been carried out. A compilation of the relevant papers is given by van Eimern et al. (1964). The reduction in wind velocity in the lee of such an installation is controlled predominantly by its porosity (Fig. 2.13) and by the angle of impact of the wind (Nägeli 1954). When investigating the windbreak effect of a double line of trees, van Eimern (1957) observed that the relative reduction in the lee of the trees is strongly dependent on the wind strength over the open ground in front of the trees. Wilson (1987) noted a slight increase in the sheltering effect of a 1.12-m-high fence when the lower part of the fence, with a porosity of 50%, is denser than the upper part.

Due to the difficulties in obtaining the relevant data, little information is available on the turbulence parameters in stands. Allen (1968) found that the turbulence intensity i_u ($i_u = \sqrt{\overline{u'^2}}/\bar{u}$) is almost independent of the height, whereas Raynor et al. (1970) observed a pronounced maximum in the canopy (Fig. 2.14). For low-plant stands, such as grain and maize, some data are available and these also show a maximum of the turbulence intensity in the upper part of the canopy (e.g., Shaw et al. 1974).

In the lee of windbreak hedges, a maximum of turbulence intensity and shear stress was noted at the level of the hedge top (Bradley and Mulhearn 1983;

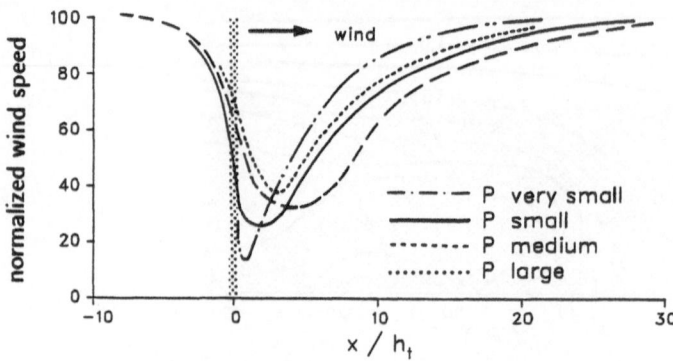

Fig. 2.13. Horizontal wind speed profiles for different porosities (in % of the value over open ground). (After Nägeli 1954)

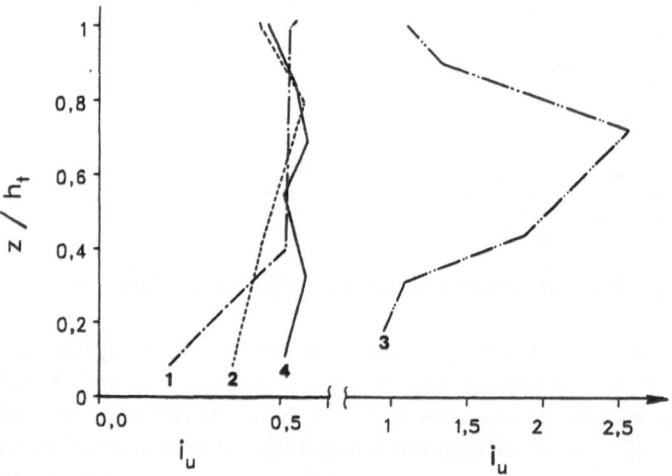

Fig. 2.14. Vertical profiles of turbulence intensity i_u *1* Atm. Sci. Res. Tech. Area; *2* Atm. Sci. Res. Tech. Area; *3* Raynor et al. (all after Cionco 1971): *4* Allan (1968)

Finnigan and Bradley 1983). This can be ascribed to the strong vertical shear which leads to a very effective transformation of kinetic energy of the mean flow into turbulent energy. Figure 2.15 illustrates the situation observed for air flow above and through a 1.2-m-high fence with 50% porosity.

2.3.2 Wind Tunnel Experiments

The greatest problem in field investigations is the selection of a suitable site. A location, which from a scientific point of view might be ideal, sometimes has to

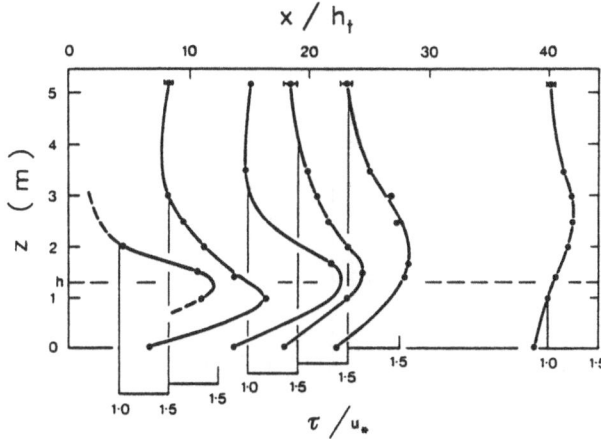

Fig. 2.15. Vertical profile of shear stress for a porous fence (normalized to friction velocity of the undisturbed velocity profile). (After Bradley and Mulhearn 1983)

be discarded for a less ideal situation as it might be inaccessible or the logistic difficulties might be insurmountable. For a number of problems, wind tunnels are a suitable substitute for field investigations, as here measurements can be carried out under controlled conditions. Such experiments are particularly useful for determining those parameters which are especially difficult to measure in field investigations.

The disadvantages of wind tunnel experiments result from difficulties encountered when taking into account thermal stratification of the atmosphere and the lack of wind-direction change with height due to the Coriolis force.

For the presentation of some wind tunnel results the coordinate system is oriented according to the direction of the superimposed wind (x-axis), while the y-axis is perpendicularly oriented, and the z-axis indicates the vertical direction.

Thom (1968) studied the forces acting on an individual leaf in a wind tunnel by measuring the effects on a leaf model, made of thin aluminium foil, of exposure to different wind speeds and to different impact angles between the leaf and the air flow. At identical wind speeds, shear on the leaf was lowest when the leaf was parallel to the wind direction. The greatest resistance was offered by the obstacle when its concave or convex side was oriented directly against the wind. The drag coefficient, c_d, of an artificial leaf at a certain inclination is defined as $c_d = F/(\varrho a V^2)$, with F being the total drag force and a the leaf area. c_d varied comparatively little for $0.5 < V < 1.5 \text{ m s}^{-1}$, but increased considerably at lower wind speeds. At $V < 0.3 \text{ m s}^{-1}$, values for c_d ranged from 0.05 to 0.55.

In laboratory tests with a natural pine branch, Grant (1983) found that c_d increases with wind velocity. The drag coefficient rose from 0.05 to 0.4 when V increased from 0.3 to 4.5 m s^{-1}. The author ascribed this situation to a change from the flow through the obstacle at low velocity to flow around the obstacle at higher velocity.

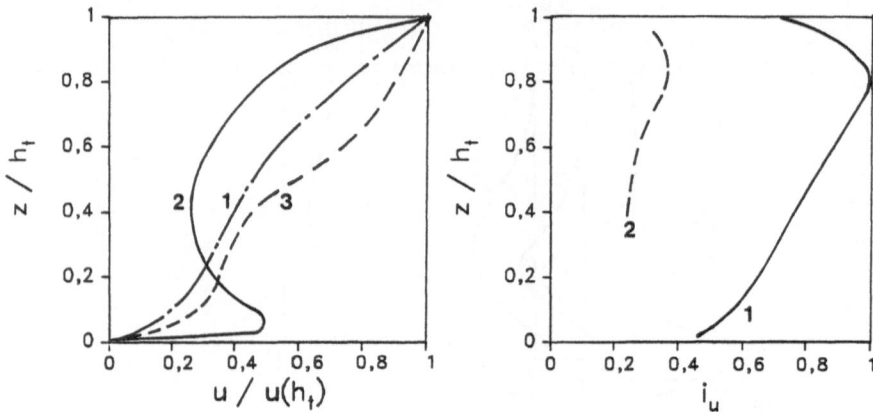

Fig. 2.16. Profiles from wind tunnel experiments. *Left* Mean wind speed: *1* Plate and Quraishi (1965); *2* Sadeh et al. (1971); *3* Finnigan and Mulhearn (1978). *Right* Turbulence intensity: *1* Sadeh et al. (1971); *2* Finnigan and Mulhearn (1978)

Plate and Quraishi (1965) investigated changes in wind velocity for a flow through an obstacle made of 10-cm-high, flexible plastic strips. They found the typical retardation encountered in canopies. The normalized velocity profiles show a similar shape for different, superimposed velocities. The authors pointed out that this effect could be the result of the shape of the obstacle.

In a similar experiment, Seginer et al. (1976) used vertically oriented 24-cm-long and 0.24-cm-thick aluminum pipes as obstacles. In an extensive test procedure, wind speed and wind shear together with the height of the obstacles were changed. In addition to the mean wind profile within and behind the obstacle, a number of turbulence profiles were obtained. While the turbulence intensity within a stand is independent of the height, the shear stress attains its maximum value in the upper portion of the canopy.

In another wind tunnel experiment, Finnigan and Mulhearn (1978) used 5-cm-high, extremely flexible plastic rods as an artificial stand, with the aim of investigating the Hanomi phenomenon, the formation and advance of wave-like movements in the upper zone of grain fields. The vertical profiles of mean wind velocity and various turbulence parameters observed by them (Fig. 2.16) are similar to those found in other laboratory experiments.

A more realistic forest model was used in the experiments of Meroney (1968). The elements of the obstacle consisted of 18-cm-high plastic trees with a trunk height of 5 cm and a crown diameter of 7 cm. For an individual model tree, a drag coefficient of $c_d = 0.72$ was found. Furthermore, the measurements show a retardation of the current on the leeward side of the obstacle. The lateral extension of this zone increases in an approximately linear fashion with increasing distance from the obstacle. At a distance of three to four crown diameters from the tree center, the velocity reduction follows approximately a Gaussian distribution (Fig. 2.17).

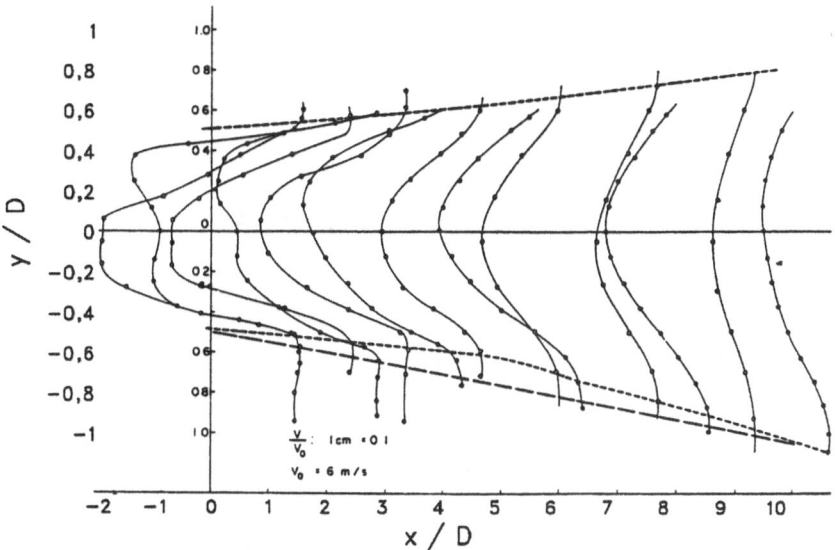

Fig. 2.17. Lateral wind speed profiles in the lee of an isolated model tree (*open circles*) and a natural tree (*solid circles*). (After Meroney 1968)

Following the tests with a single model tree, flow in and above a stand made up of such model trees was investigated. After a horizontal homogeneous inflow zone, above which a power profile of the wind had established itself, the model forest began with a fairly sharp edge. It was found that the velocity is already reduced at some distance ahead of the forest, the degree of reduction increasing as the wind penetrates the stand. Within the trunk zone there is initially a secondary velocity maximum; however, this is then reduced farther into the forest (Fig. 2.18, $x = 10$ m).

In the same experiment, the turbulence intensity in front of the stand showed little variation with height. Within the stand itself a pronounced maximum was found in the upper part of the canopy (Fig. 2.19). At a distance of $x = 4$ m, corresponding to about 20 times the canopy height, a new equilibrium for wind and turbulence is established.

Sadeh et al. (1971) used the same wind tunnel for their experiments as that used by Meroney (1968); they used a similar plastic tree model, with the crown diameter increased to 8 cm. The other geometrical parameters were the same as those used by Meroney (1968). As a result of the similarity of the experimental arrangement and the individual obstacles, the vertical profiles of wind and turbulence intensity differed only a little from those illustrated in Figs. 2.18 and 2.19. The horizontal distribution of the wind velocity within and above the stand shows a nearly exponential decrease as has already been found in field investigations (Fig. 2.20).

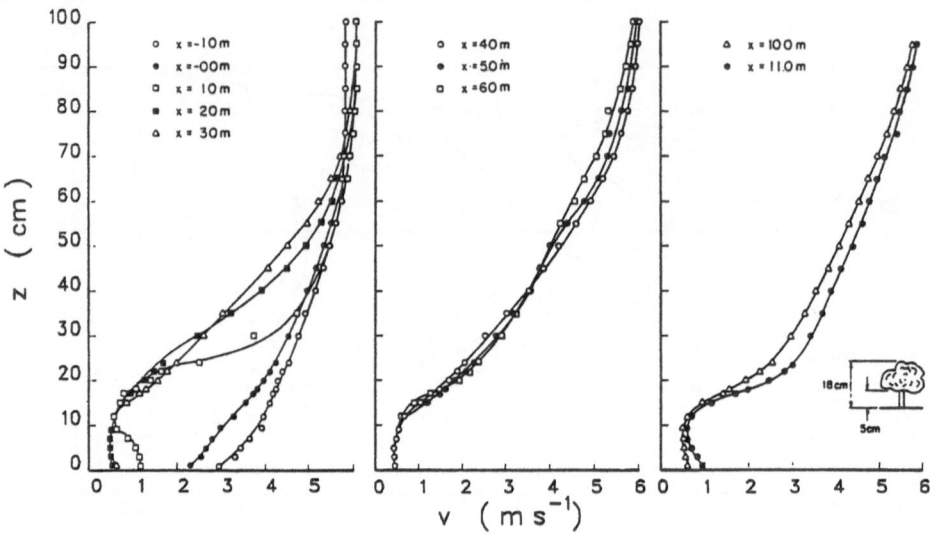

Fig. 2.18. Vertical profiles of wind speed at different locations within a stand. (After Meroney 1968)

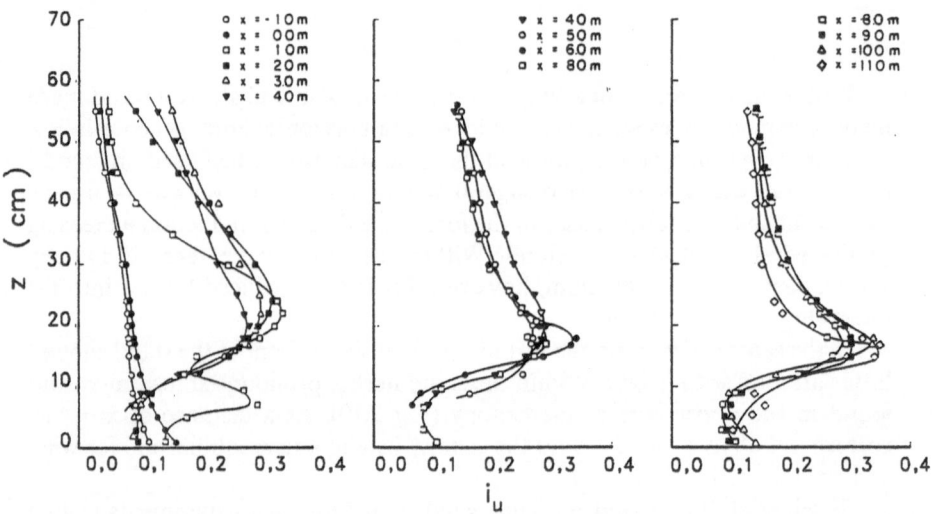

Fig. 2.19. Vertical profiles of turbulence intensity at different positions within a stand. (After Meroney 1968)

Ruck and Schmitt (1986a) investigated the wind and turbulence distribution around two individual trees, one being an 18-cm-high solid model cone and the other a 33-cm-high natural sugarloaf spruce. The air flow was measured by laser-Doppler methods which permit the joint determination of mean velocities and turbulence.

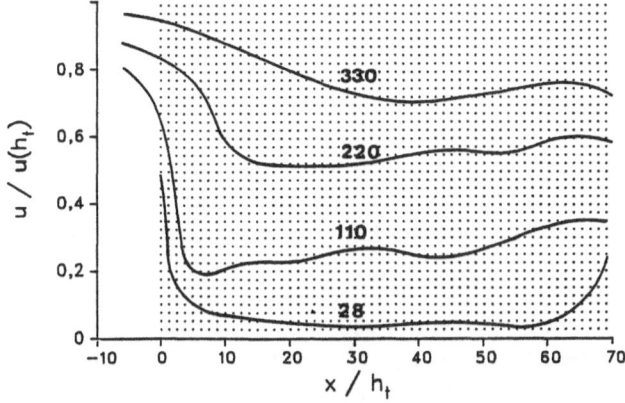

Fig. 2.20. Horizontal profiles of wind speed in and above an artificial obstacle. (After Sadeh et al. 1971)

For the impermeable model cone with and without a 5-cm-high trunk, the results were as presented in Fig. 2.21. In the lee of the obstacle there is a zone of reverse flow that is approximately one crown diameter long. For the model tree with a trunk, strong flow under the obstacle occurs, and at some distance behind the tree a velocity minimum was encountered close to the ground. Although the trunk increases the volume of the tree (defined as the space filled out by the canopy elements) only marginally, it causes a pronounced difference in the wind pattern. Especially when underflow is permitted, the disturbance is stronger and decreases slowly farther downwind.

In the second set of measurements Ruck and Schmitt (1986a) used a sugarloaf spruce with a volume porosity of $P = 0.93$ as an obstacle in the wind tunnel. The

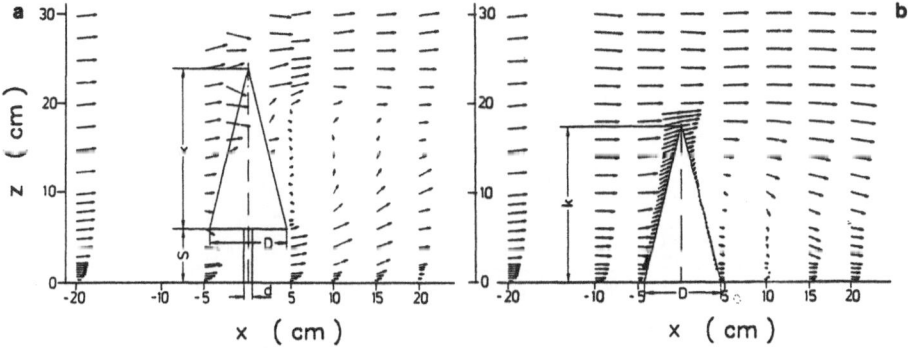

Fig. 2.21. Velocity pattern around a model cone **a** with trunk and **b** without. (After Ruck and Schmitt 1986a)

crown, with a diameter of about 23 cm, could still be penetrated at an undisturbed velocity of 1.2 m s^{-1}, despite a strong retardation of the air flow.

At the same time there was strong overflow which leads to a zone of reverse
flow some distance behind the obstacle, also about one crown diameter long.
When underflow was permitted, a velocity maximum close to the ground was
observed together with an elevated zone of reverse flow (Fig. 2.22).

The fluctuations parallel and vertical to the direction of the oncoming wind
were determined as turbulence parameters. These two quantities are shown in
Fig. 2.23 normalized against the undisturbed velocity. Both parameters show
pronounced maxima in the lee of the obstacle, with a decrease in size with
distance from the obstacle and, at the same, a shift to higher levels. The structure
of vertical profiles thus measured shows good agreement with the results of
other wind tunnel experiments and field observations.

Ruck and Schmitt (1986a) also measured the wind fields around complete
stands. However, in contrast to e.g., Meroney (1968) they did not arrange the

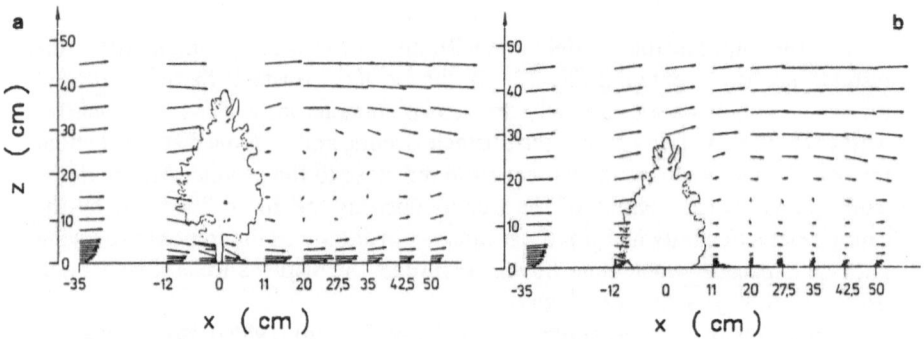

Fig. 2.22. Velocity pattern around a live sugar loaf spruce **a** with trunk and **b** without.
(After Ruck and Schmitt 1986a)

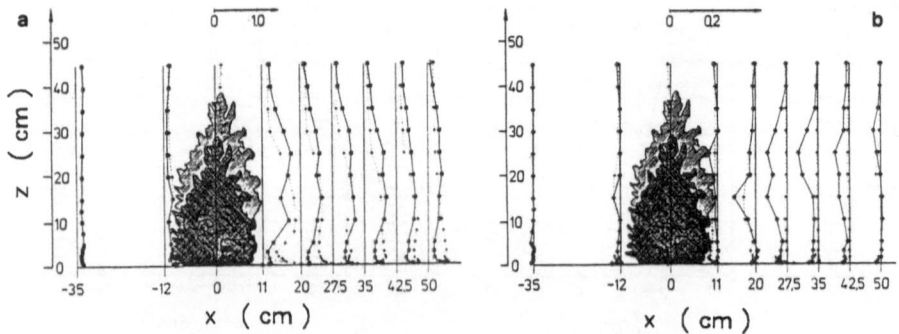

Fig. 2.23. Observed turbulence for flow around a sugar loaf spruce. (After Ruck and
Schmitt 1986a), **a** $i_u/\bar{u}(H)^2$; **b** $\tau/\bar{u}(H)^2$

individual obstacles at random but in rows. This leads to a pronounced channeling effect in the gaps within the stand, whereas in the tree zone a strong retardation was found.

2.3.3 Analytic Description and Numerical Models

The first analytic description of air flow in a stand was based on the assumption that turbulent transport in a canopy may be treated in the same way as over open ground, provided that dissipation of momentum on the stand elements is permitted. Furthermore, it was assumed that the drag force of the canopy elements is proportional to the square of the velocity. In the stationary, horizontally homogeneous case this drag force is equal to the divergence of the momentum transport. Assuming in addition, that the momentum transport may be described by a flux gradient approach and that the turbulent diffusion coefficient can be determined by a mixing length relationship, the problem is reduced to the solution of the following system of equations from the ground up to the stand height h_t:

$$\frac{d\tau}{dz} = \varrho c_d b(z) u^2,$$

$$\tau(z) = \varrho K \frac{du}{dz},$$

$$K = l^2 \frac{du}{dz}. \tag{2.9}$$

In these equations, $\tau(z)$ is the shear stress at height z, ϱ the air density, c_d a drag coefficient of the stand elements, $b(z)$ the vertical distribution of the leaf area density, u the wind speed, K the turbulent diffusion coefficient for momentum and l the mixing length.

Provided that c_d and l are constant, Cionco (1962, 1965) and Inoue (1963) arrived at

$$u(z) = u(h_t) e^{\alpha_t(z/h_t - 1)} \tag{2.10}$$

as the solution of the system of equations [Eq. (2.9)] with the constraints of $u(z = h_t) = u(h_t)$. The canopy index α_t is obtained from

$$\alpha_t = \left[0.5 c_d b(z) \frac{h_t^3}{l^2} \right]^{1/3}. \tag{2.11}$$

Despite the partly unrealistic assumptions leading to this exponential wind profile, Eq. (2.10) is frequently applied. The reason for this is that it results in rather good agreement with observations, provided that a suitable canopy index is chosen. The values of α_t, compiled by Cionco (1972) for different types of vegetation, characteristically range from two to three.

If a constant diffusion coefficient K is used instead of $l = $ constant, the vertical distribution of the velocity, according to Thom (1971) may be written as:

$$u(z) = u(h_t)\left[1 + \alpha_{t2}\left(1 - \frac{z}{h_t}\right)\right]^{-2}. \tag{2.12}$$

For a given mixing length (varying with height), Cowan (1968) found the following wind profile:

$$u(z) = u(h_t)\left[\frac{\sinh(\alpha_{t3} z/h_t)}{\sinh(\alpha_{t3})}\right]^{0.5}, \tag{2.13}$$

with $u(z = 0) = 0$ and $u(z = h_t) = u(h_t)$.

The parameters α_{t2} and α_{t3} appearing in Eqs. (2.12) and (2.13) represent empirical coefficients which may be obtained from a combination of the parameters in Eq. (2.11).

Only Eq. (2.13) fulfills both boundary conditions at ground level and at $z = h_t$, whereas Eqs. (2.10)–(2.12) result in a correct solution only for $z = h_t$, showing a deficiency at $z = 0$. With suitable selected empirical coefficients, however, all three equations lead to similar distributions of the wind profile in the stand (Fig. 2.24).

Despite the similarity of the various profiles for u, the corresponding profiles of K differ considerably from one another. For Eqs. (2.10) and (2.12) the diffusion coefficients may be expressed by an exponential function, according to Uchijiama (1962), whereas in Eq. (2.13) $K = $ constant was assumed. This led Raupach and Thom (1981) to conclude that for any reasonable assumed value of K, a plausible wind profile is obtained.

The simple models described above are based on the assumption that for turbulent flow in a stand, the flux-gradient relationship is applicable. However, wind tunnel experiments and field investigations show that this approach is frequently inapplicable and that turbulent fluxes against the gradient of the mean values may be encountered (Raupach and Legg 1984; Denmead and

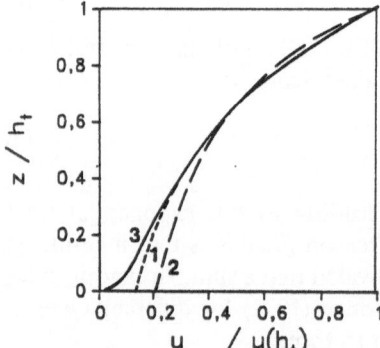

Fig. 2.24. Comparison of analytic wind profiles (after Raupach and Thom 1981). *1* With $\alpha_t = 2$ in Eq. (2.10); *2* with $\alpha_{t2} = 1.2$ in Eq. (2.12); *3* with $\alpha_{t3} = 4$ in Eq. (2.13)

Bradley 1985). The latter two authors ascribed this mainly to the sporadic intrusion of large eddies into the stand.

Prior to these experimental results, models were already being developed to describe turbulent flow by equations for the second-order moments. Wilson and Shaw (1977) proposed that pressure forces account for the majority of the total drag of a stand. Their model of a neutral stratified atmosphere is based on a coupled set of equations for the mean wind speed, the shear stress and the variances of the three wind components. The velocity profile thus obtained shows a small maximum near the ground, indicating that transport of momentum takes place against the gradient. The calculated profile of $\tau(z)$ also shows the expected shape (Fig. 2.25).

In the model of Lewellen et al. (1979), in addition to these equations those for the turbulent heat flux and other passive components can also be solved. Above high trees the wind profile in a neutral stratified atmosphere can be described by the logarithmic profile:

$$u(z) = \frac{u_*}{\kappa} \ln \frac{z - d}{z_0}. \tag{2.14}$$

In this expression, u_* is the friction velocity, κ the von Karman constant, d the zero displacement and z_0 the roughness length. A similar logarithmic law may be assumed for temperature and humidity profiles.

The zero displacement is usually the height, to which the wind profile calculated from Eq. (2.14) has to be shifted in order to obtain optimal agreement with the observed profile. Thom (1971) and Jackson (1981) interpreted d as the mean height at which momentum is absorbed by the canopy. Another problem related to the zero displacement is whether or not identical values for d may be used for momentum, heat and humidity. Hicks et al. (1979) had actually postulated that additional lengths would be required for heat and humidity.

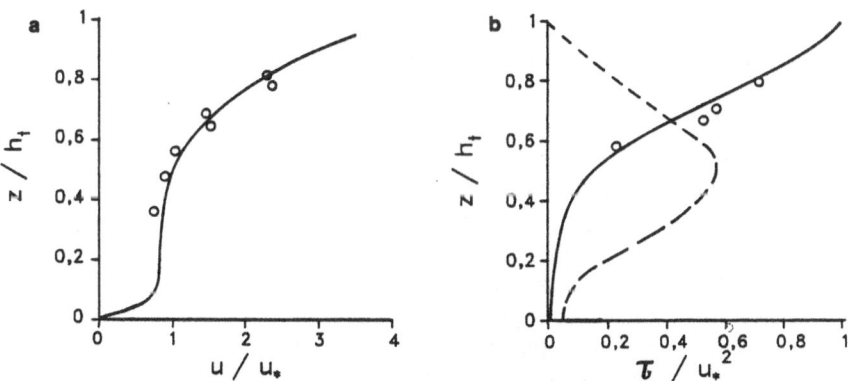

Fig. 2.25. Dimensionless profiles of **a** velocity and **b** shear stress (after Wilson and Shaw 1977). *Dashed line* Leaf area density; *open circles* observed velocity and shear stress

For practical purposes d and the roughness length z_0 are given in fractions of the stand height. Cionco (1985) proposed $d = 0.7 h_t$ and $z_0 = 0.14 h_t$.

As compiled by Kurata (1982), other authors report the following values: $0.05 < z_0/h_t < 0.2$ and $0.70 < d/h_t < 0.80$.

The flow pattern within a forest is controlled by large-scale weather conditions, season and time of day. In the opposite direction, tall vegetation influences large-scale wind patterns, temperature and humidity distribution through the exchange of momentum and heat. In order to obtain a realistic model of the conditions within a canopy, many of the potential interactions between the stand and the individual meteorological variables have to be taken into account. Deardorff (1978) developed a simple scheme for the parameterization of the soil temperature including a layer of vegetation. It contains the solution of an abbreviated energy budget equation for maintaining the temperature of a representative stand element. This, in turn, allows the determination of temperature and humidity within the vegetation layer. By varying the density of the plant cover, n_c, the equation permits one to simulate conditions of unforested ($n_c = 0$) and forested ground ($n_c > 0$). Although at $n_c = 1$ the diurnal surface temperature variation is strongly attenuated, it is still quite notable as the exchange of energy and humidity between soil and vegetation continues. Deardorff (1978) found rather good agreement between his results for $n_c = 0.75$ and the observations of Penman and Long (1960).

McCumber (1980) combined Deardorff's model with a one-dimensional boundary layer model. By using actual meteorological data, and taking the effects of vegetation on meteorological variables into account, he could show that there were temperature differences of up to $3°C$ and that the height of the mixing layer differed by up to 300 m. Diurnal temperature variations of stand elements, the vegetation layer and the surface, calculated in such a simulation, are illustrated in Fig. 2.26.

Garratt (1983) applied the same canopy parameterization in his numerical investigations of cold air drainage winds in a one-dimensional model. At $n_c = 0.5$ he observed a reduction in wind speed by about 50% which he ascribed to the increased friction effect exerted by the assumed 15-m-high stand. Furthermore, he found that the velocity maximum was elevated from ground level to about twice the tree height.

Yamada (1982) included in his model the influence of tall stands by using the approach of Wilson and Shaw (1977). With an assumed tree height of 20 m, he calculated temperature, humidity, wind speed and turbulence for 11 levels within the vegetation layer. The results of his one-dimensional simulations represent the characteristic phenomena observed in a forest quite well. In addition to the low wind speed, the simulations also show the unstable thermal stratification close to the ground during the night and the more stable conditions during the day. This applies, however, only for very large values of n_c. As observed in nature, the calculated turbulence, expressed by the turbulent kinetic energy E, also shows a maximum within the canopy (Fig. 2.27). In this model the effect of evapotranspiration by the plants is not considered, although, as shown

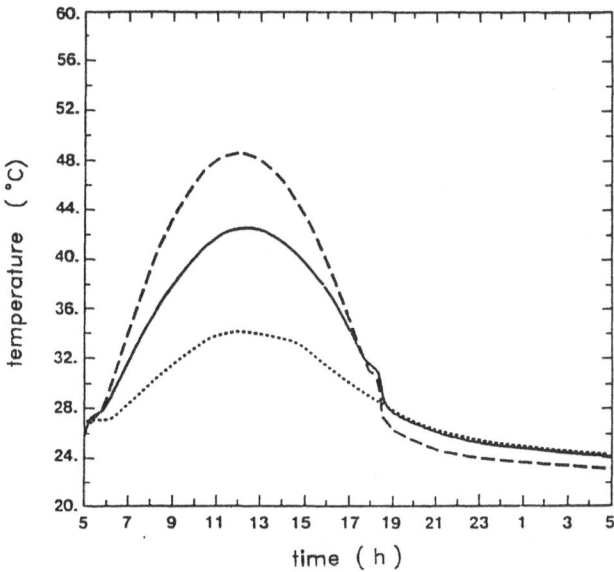

Fig. 2.26. Simulated diurnal temperature variation. *Dotted line* Surface; *solid line* vegetation layer; *dashed line* stand elements. (After McCumber 1980)

by Koch (1987), evapotranspiration can significantly affect humidity and temperature patterns, especially during the daytime.

By combining a mesoscale wind model with a canopy model, Cionco (1985) attempted to take into account the vegetation cover in multidimensional simulation models. By applying a variation method, he determined diagnostically the wind field near the ground under the influence of topographic and thermal effects. He assumed that over each point of his area of calculation, the logarithmic profile for wind would be valid. Using a coupling parameter which is available from a table for a variety of types of vegetation, Cionco (1985) simulated the wind velocities in the upper part of the canopy from the previously calculated flow. By assuming an exponential profile of wind speed within the stand [Eq. (2.10)], he found that the three-dimensional velocity distribution was affected by topography and vegetation. A disadvantage of his model is the rather rough estimation of the mesoscale wind field above the stand.

Yamada (1985) removed this shortcoming by expanding his one-dimensional model to three dimensions. He investigated the structure of the nocturnal wind system, temperature and turbulence in the California geyser area. The velocities of the nocturnal cold air drainage flows obtained agreed well with the observed data. Calculations of the dispersal of air pollutants also showed satisfactory agreement with the observations. This was ascribed in part by Yamada (1985) to the fact that the turbulence could be simulated in the right order of magnitude. In particular, the production by wind shear and the generation by the stand

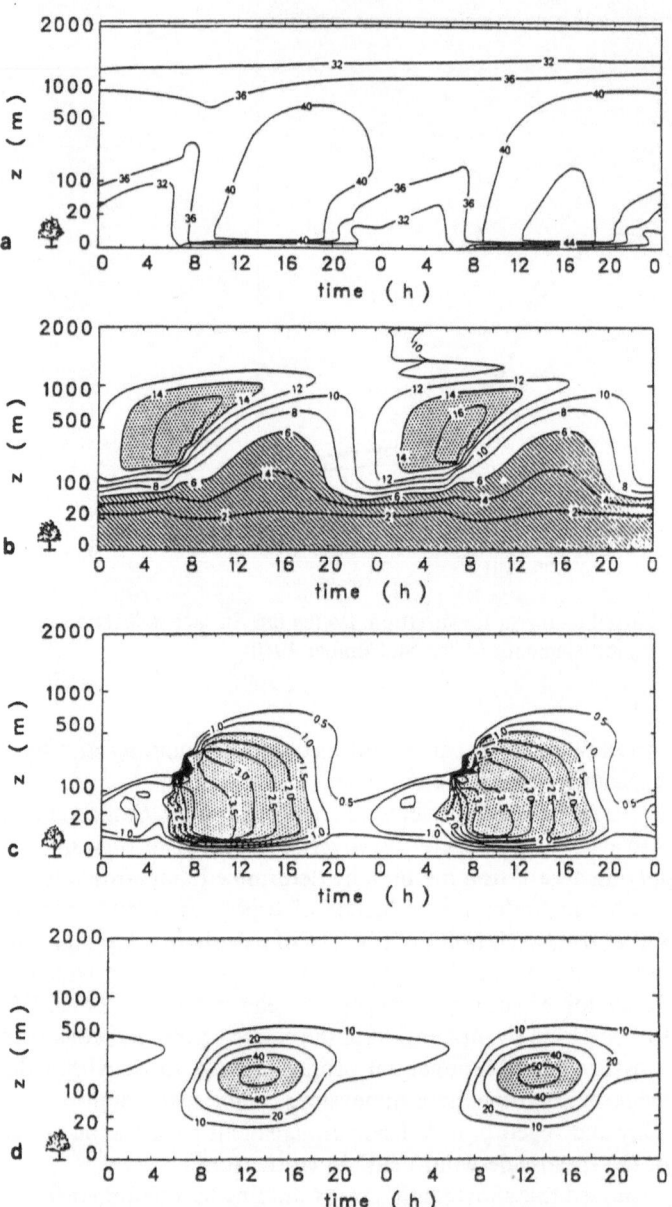

Fig. 2.27. Simulated diurnal variation of **a** potential temperature at $n_c = 1.0$; **b** wind speed at $n_c = 0.9$; **c** turbulent kinetic energy at $n_c = 0.9$; and **d** turbulent diffusion coefficient at $n_c = 0.9$ (After Yamada 1982)

elements led to larger diffusion coefficients, which allowed more realistic dispersal calculations.

The influence of small-scale deforestation on nocturnal drainage flows and on the local climate was investigated by Gross (1987) by three-dimensional numerical simulation. Under otherwise equal conditions, calculations were carried out for cases with and without forest cover. The comparison of the results of the two calculations allowed certain conclusions to be drawn and these will be described in Chapter 4.

3 Air Flow Around and Through Individual Trees

In this chapter, a numerical model will be introduced to permit the calculation of the distribution in time and space of wind, turbulence, temperature and humidity in and around individual trees.

3.1. The Numerical Model

3.1.1 The System of Equations

The system of equations, on which this model is based, consists of a set of conservation equations for momentum, mass and energy. Neglecting the Coriolis force, the Boussinesq-approximated Navier–Stokes equation, the continuity equation and the first law of thermodynamics may be expressed as follows:

$$\frac{\partial u_i}{\partial t} + u_j \frac{\partial u_i}{\partial x_j} = -\frac{1}{\varrho}\frac{\partial p}{\partial x_i} + g\frac{\theta}{\bar{\theta}}\delta_{i3} + v_u \frac{\partial^2 u_i}{\partial x_j^2}, \tag{3.1}$$

$$\frac{\partial u_i}{\partial x_i} = 0, \tag{3.2}$$

$$\frac{\partial \theta}{\partial t} + u_j \frac{\partial \theta}{\partial x_j} + w\frac{\partial \tilde{\theta}}{\partial z}\delta_{j3} = v_\mathrm{T}\frac{\partial^2 \theta}{\partial x_j^2}. \tag{3.3}$$

The symbols used are: u_i for the velocity component in the direction x_i, ϱ for the air density (taken as constant), p and θ for the deviations of pressure and potential temperature, respectively, from a basic state $(\tilde{p}, \tilde{\theta})$, v_u for the kinematic viscosity and v_T for the molecular diffusivity of temperature.

If one of the dependent variables in the above equations is designated as ϕ, this may be split into a mean value $\bar{\phi}$ and a deviation ϕ' from it. In this, $\bar{\phi}$ represents the mean over a time interval and a certain volume. In a numerical model this time interval is represented by the time step Δt. The volume is determined from the width of the grid units in the three directions Δx, Δy and Δz.

Splitting up the variables and then averaging them, results in

$$\frac{\partial \bar{u}_i}{\partial t} + \bar{u}_j\frac{\partial \bar{u}_i}{\partial x_j} + \frac{\overline{\partial u'_i u'_j}}{\partial x_j} = -\frac{1}{\varrho}\frac{\partial \bar{p}}{\partial x_i} - \frac{1}{\varrho}\frac{\overline{\partial p'}}{\partial x_i} + g\frac{\bar{\theta}}{\bar{\theta}}\delta_{i3} + v_u \frac{\partial^2 \bar{u}_i}{\partial x_j^2} + v_u\frac{\overline{\partial^2 u'_i}}{\partial x_j^2}. \tag{3.4}$$

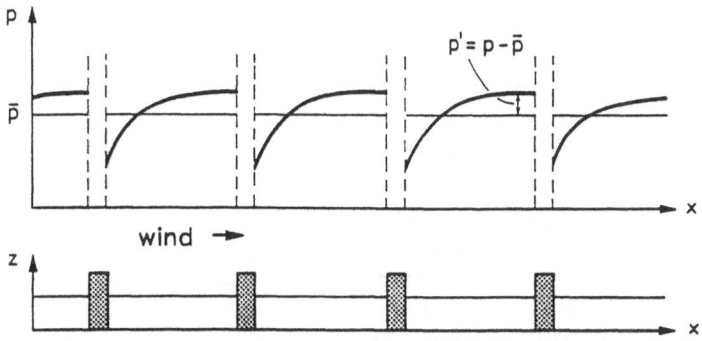

Fig. 3.1. Schematic pressure pattern around impermeable fences oriented perpendicular to the wind direction

The second and fifth terms on the right-hand side will not appear when averaging and the differentiation in space are commutative. In meteorology, this is usually the case, but it is not so within a stand where the averaged volume is partly occupied by stand elements. This will be explained in the following example.

The pressure field resulting from air flow over a row of fences at right angles to the wind is shown in Fig. 3.1. Between two subsequent obstacles there is $\partial p'/\partial x > 0$ and thus, horizontal averaging results in $\overline{\partial p'/\partial x} > 0$. As by definition there is $\partial \overline{p'}/\partial x = 0$, it follows directly that in this case differentiation in space and averaging are not commutative. Therefore, the second term on the right-hand side of Eq. (3.4) represents the sum of the different pressure forces on the windward and leeward sides of the stand elements and, consequently, has to appear in this equation.

A similar consideration leads to the conclusion that the fifth term on the right-hand side also differs from zero. It may be interpreted as the mean friction force of the plants.

Applying the rules of averaging to Eqs. (3.2) and (3.3) leads to

$$\frac{\partial \bar{u}_i}{\partial x_i} = 0, \tag{3.5}$$

$$\frac{\partial \bar{\theta}}{\partial t} + \bar{u}_j \frac{\partial \bar{\theta}}{\partial x_j} + w \frac{\partial \tilde{\theta}}{\partial z} \delta_{j3} + \frac{\overline{\partial u'_j \theta'}}{\partial x_j} = \nu_T \frac{\partial^2 \bar{\theta}}{\partial x_j^2}. \tag{3.6}$$

The correlation products $\overline{u'_i u'_j}$ and $\overline{u'_j \theta'}$ included in the equations can be related to the mean variables. Although it is possible to develop a separate system of equations for these products also, this would be rather laborious in a three-dimensional case. Consequently, the flux-gradient approach will be used.

The correlation products of the velocity fluctuation may be written as

$$-\overline{u'_i u'_j} = K_m \left(\frac{\partial \bar{u}_i}{\partial x_j} + \frac{\partial \bar{u}_j}{\partial x_i} \right) - \frac{2}{3} E \delta_{ij}. \tag{3.7}$$

Here, K_m is the turbulent diffusion coefficient of momentum and E the turbulent kinetic energy ($E = 0.5\overline{u_i'^2}$). The term containing E is necessary if Eq. (3.7) is to be used for normal Reynolds stress too. Without it, the sum of the normal Reynolds stress would be zero due to the applicability of the continuity equation [Eq. (3.5)]. It has, however, to be equivalent to twice the turbulent kinetic energy.

In the numerical model presented here the turbulent fluxes are treated according to Eq. (3.7) although the applicability of this approach in a stand has been questioned by, e.g., Raupach and Thom (1981), Raupach and Legg (1984) and Denmead and Bradley (1985). It has to be remembered, however, that these authors came to this conclusion by comparing observations with the results of one-dimensional model calculations. One-dimensional in this context refers to horizontal homogeneity, changes being allowed only in the vertical direction. Furthermore, it implies that any advection which may also be directed against the gradient of the mean quantity is not taken into account. As advective processes are of greater importance than diffusive processes in the simulations carried out here, the application of the flux-gradient approach appears to be admissible.

The terms which represent the effects of the stands in the model equations are treated according to Wilson and Shaw (1977). Consequently, the viscous drag is neglected in favor of the form drag. The latter is expressed as

$$\frac{1}{\varrho}\frac{\overline{\partial p'}}{\partial x_i} = c_d b \bar{u}_i V. \tag{3.8}$$

When molecular diffusion is neglected in favor of turbulent diffusion, the system of equations may be written as

$$\frac{\partial \bar{u}_i}{\partial t} + \bar{u}_j \frac{\partial \bar{u}_i}{\partial x_j} = -\frac{1}{\varrho}\frac{\partial \bar{p}}{\partial x_i} - c_d b \bar{u}_i V + g\frac{\bar{\theta}}{\tilde{\theta}}\delta_{i3}$$

$$+ \frac{\partial}{\partial x_j}\left[K_m\left(\frac{\partial \bar{u}_i}{\partial x_j} + \frac{\partial \bar{u}_j}{\partial x_i}\right) - \frac{2}{3}E\delta_{ij} \right], \tag{3.9}$$

$$\frac{\partial \bar{u}_i}{\partial x_i} = 0, \tag{3.10}$$

$$\frac{\partial \bar{\theta}}{\partial t} + \bar{u}_j \frac{\partial \bar{\theta}}{\partial x_j} + w\frac{\partial \tilde{\theta}}{\partial z}\delta_{j3} = \frac{\partial}{\partial x_j}\left(K_h\frac{\partial \bar{\theta}}{\partial x_j} \right). \tag{3.11}$$

The turbulent diffusion coefficient here is expressed through the Prandtl–Kolmogorov relation

$$K_m = al\sqrt{E}. \tag{3.12}$$

For the length scale l, the mixing length approach according to Blackadar (1962) is used: $l = \kappa z/(1 + \kappa z/l_\infty)$; $z \geq z_0$. In order to maintain the turbulent kinetic energy, a prognostic equation is solved. This is found by subtracting Eq. (3.4) from eq. (3.1), multiplying the result by u_i' and then averaging it. This leads to

$$\frac{\partial E}{\partial t} + \bar{u}_j \frac{\partial E}{\partial x_j} = \frac{\partial}{\partial x_j}\left(K_m \frac{\partial E}{\partial x_j}\right) + K_m\left(\frac{\partial \bar{u}_i}{\partial x_j} + \frac{\partial \bar{u}_j}{\partial x_i}\right)\frac{\partial \bar{u}_i}{\partial x_j}$$

$$- K_h \frac{g}{\bar{\theta}}\frac{\partial\bar{\theta}}{\partial x_j}\delta_{j3} - \frac{E^{1.5}}{a^{-3}l} + c_d b V^3. \tag{3.13}$$

In this equation certain additional assumptions have been made for the determination of unknown correlation products (e.g., Rodi 1980) and the flux-gradient relationship was applied again.

For the stratification-dependent factor a in Eqs. (3.12) and (3.13) the approach:

$$a = \frac{a_0}{\Phi} \tag{3.14}$$

is used with $a_0 = 0.45$, while $\Phi(z/L_*)$ represents the local profile function which may, according to Businger et al. (1971), be expressed as:

$$\Phi = 1 + 4.7\frac{z}{L_*} \quad \text{for} \quad \frac{\partial\theta}{\partial z} > 0, \tag{3.15}$$

$$\Phi = \left(1 - 15\frac{z}{L_*}\right)^{-0.25} \quad \text{for} \quad \frac{\partial\theta}{\partial z} < 0. \tag{3.16}$$

Here, L_* represents the Monin–Obukhov stability length. The diffusion coefficient for heat, K_h, is related to that for momentum through the turbulent Prandtl number, for which a value of 1.35 is used.

Kurz (1977) found a relationship between horizontal and vertical diffusion by comparing the standard deviation of the concentration in the lee of a continuous point source. By this method, the horizontal diffusion at right angles to the flow direction at neutral thermal stratification is 2.5 times the vertical diffusion. This factor is used regardless of the local thermal stratification which is actually developing.

To determine the pressure disturbance, a Poisson equation is solved where, in addition, the continuity equation is used. The pressure distribution thus obtained assures that the velocity field fulfills both the equation of motion and the continuity equation. For the solution, the algorithm of Schumann and Volkert (1984) is used.

3.1.2 Boundary Values and Initial Conditions

The system of equations described above is solved for a limited area. Consequently, the variables or their derivatives along the boundaries of the area have to be known. Values for these variables along the boundaries must either be acquired by observation or be derived from additional physical relations.

The area of integration is usually a rectangular-based volume. Only the lower boundary, the ground, is physically defined, while the other boundaries are fixed more or less arbitrarily. The boundaries should be as far removed as possible from the center of the actual area of investigation in order to minimize their disturbing influence on the solution. Oliger and Sandström (1976) and Gal-Chen and Sommerville (1975) presented a number of combinations of possible boundary conditions.

The lower boundary, the ground, represents an impermeable rigid wall for the air flow. By additionally assuming a noslip condition along the boundary, the three velocity components are set at zero. Temperature is kept constant over the period of integration and thus $\bar{\theta}$ also vanishes. The Poisson equation for determining $\bar{p}$ is solved by setting Neumann conditions. The turbulent kinetic energy close to the ground is found from Eq. (3.12) and the relations in the Prandtl layer are found with $E = u_*^2/a_0$.

The upper boundary of the integration area at height H is considered to be a wall, through which, as through the ground, no flux can take place. This immediately results in $\bar{w} = 0$, while $\bar{u}(H)$ and $\bar{v}(H)$ may be set separately. Temperature disturbance and the vertical derivative of E disappear but $\partial \bar{p}/\partial z = 0$ is also set. This selection of boundary conditions implies that disturbances originating within the area of computation have dissipated before reaching the upper boundary.

The lateral boundaries may be considered as either open or closed. Boundaries are closed when the distribution of the meteorological variables is not influenced by what is happening within the area itself. Such constraints are used for boundaries along which inflow takes place. They may be determined accurately because the x-axis of the coordinate system is always oriented in the direction of the wind at height H. The vertical profiles of $\bar{u}$, $\bar{v}$, E and $\tilde{\theta}$ at the inflow boundaries are set or calculated, while $\bar{w}$, $\bar{\theta}$ and $\partial \bar{p}/\partial x$ are set to zero.

It is assumed that disturbances originating within the area and advancing to its outflow boundary may leave the area unimpeded. The radiation condition of Orlanski (1976) is used, as it reduces the reflection along the boundaries to a minimum. In this case, the equation

$$\frac{\partial \phi}{\partial t} = - c \frac{\partial \phi}{\partial x} \tag{3.17}$$

is solved at the boundary for a variable ϕ, c is the group velocity. This is also obtained from the above equation as the mean of the last four points of the numerical grid. The group velocity must fulfill the condition $0 < c < \Delta x/\Delta t$. For negative values, c is taken as zero. Other methods for the determination of c are given in Klemp and Lilly (1978), Miller and Thorpe (1981) and Hack and Schubert (1981).

For the pressure along the outflow boundary, a Neumann condition is used.

The shape of the stand used as an obstacle in the simulation is taken as of rotational symmetry. Because of this the flow field is calculated for only one half of the obstacle, whereas the other half is taken as the mirror image of the calculated one. It is assumed that along the symmetry plane the y-derivatives of $\bar{u}$, $\bar{w}$, E, $\bar{\theta}$ and $\bar{p}$ will disappear and that $\bar{v}$ is effectively zero.

As this system of equations also poses a problem of an initial value, initial fields for the variables have to be given to time $t = 0$. This is achieved by the solution of the one-dimensional model equations applying to horizontally homogeneous conditions. With a characteristic roughness length, z_0, and a given basic state for the temperature, a profile for the turbulent kinetic energy and a nearly logarithmic vertical profile for $\bar{u}$ are obtained. The variables $\bar{v}$, $\bar{w}$, $\bar{p}$ and $\bar{\theta}$ at the beginning of the simulation are all taken as zero.

3.1.3 Discretization and Numerical Solution

To find an exact solution for the system of equations used usually presents extreme mathematical problems, particularly because of the nonlinear character of these equations. Only for a few special cases can exact solutions be given (e.g. Schlichting 1958).

However, numerical methods usually have to be applied in order to solve the set of nonlinear partial differential equations simultaneously in a discrete manner on a numerical grid. For this purpose the integration region is split into small volumes of a certain size, $G = \Delta x \Delta y \Delta z$. Only one value of a variable, considered as representative for this volume, is then calculated. This approach corresponds to the introduction of mean volumes as outlined in Section 3.1.1.

The arrangement of dependent variables on such a numerical grid is presented in Fig. 3.2. In the center of the volume the pressure p, the diffusion coefficients K_m and K_h and the turbulent kinetic energy are all defined, whereas the respective normal velocity components are arranged along the sides. The temperature is defined at the same grid points as the vertical velocity. The grid used is of the staggered type in which the variables are displaced against each

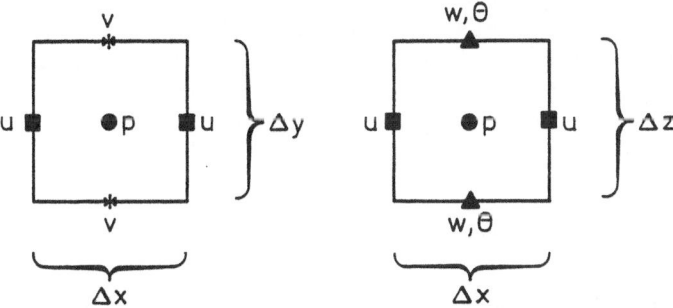

Fig. 3.2. Arrangement of the variables used on the numerical grid

other in the vertical and horizontal directions by half a grid length. This arrangement has the advantage that especially the pressure gradient term in Eq. (3.9) and the continuity equation [Eq. (3.10)] may be calculated with better numerical accuracy. These two terms must be treated as exactly as possible, because they control the balancing of the pressure and the velocity field. The above volumes are arranged within the integration region in such a way that the respective normal velocity component is situated along the boundary planes. The ground thus coincides with the grid point along which the vertical velocity is defined.

However, the disadvantage of using such a staggered grid is that each variable is not defined at the same grid point. If, for example, in the calculation of the vertical advection of $\bar{u}$, the vertical velocity $\bar{w}$ is used, it first has to be obtained through averaging over the surrounding points.

From the techniques available (finite differences, finite elements and spectral methods) for the calculation of spatial derivatives, the finite difference method is used; this is most widely employed in mesoscale modeling.

The differential terms here are approximated by centered differences. Other methods such as the upwind differencing possess a large numerical diffusion and thus may distort the calculated fields. However, occasionally, an artificial diffusion is introduced in order to keep the model calculations numerically stable. The decision as to which method is used mainly depends on the problem being investigated.

For integration into time, the Adams–Bashforth method is used, as described by Roache (1982). It is an explicit method in which the variables of two immediately preceeding time steps are required.

In this time integration, the time step Δt may not be chosen as arbitrarily large because, at least for the linear problem, the Courant–Fridrichs–Lewy criterion must be fulfilled. Due to the lack of an alternative, the time increment is calculated also for the nonlinear case from

$$\Delta t < \left[\frac{u}{\Delta x} + \frac{v}{\Delta y} + \frac{w}{\Delta z} + \frac{2K_{hh}}{(\Delta x)^2} + \frac{2K_{hh}}{(\Delta y)^2} + \frac{2K_{hv}}{(\Delta z)^2} \right]^{-1} \tag{3.18}$$

(Roache 1982; Tangermann-Dlugi 1982).

At each grid point in the computation area, Δt is calculated according to the local flow and turbulence conditions. The overall smallest time increment is used for the advance in time. When using typical values for $\bar{u}$ and Δx ($\bar{u} = 2 \text{ m s}^{-1}$, $\Delta x = 1 \text{ m}$), Δt is found to range from 0.1 to 0.5 s.

3.1.4 Parameterization of the Stand in the Model

The distribution of the meteorological variables in a stand is controlled especially by the leaf area density LAD, $b(z)$. This is defined as the leaf area in m^2 per m^3 of stand volume and characterizes the degree of filling of the volume by leaves, branches and twigs. An additional parameter is obtained through vertical integration of LAD. The resulting leaf area index (LAI) is given as m^2 leaf area per m^2 ground surface and represents a dimensionless measure for the degree of cover. At an LAI of 7, which is typical for conifers, the area of projection of the tree onto the ground is covered by a sevenfold larger leaf area.

For the numerical simulation, the three-dimensional distribution of LAD is required; however, this is not usually available. In the literature, mostly vertical profiles of LAD are presented, from which $b(x, y, z)$ has to be reconstructed by assuming a certain tree shape and an equal distribution of the stand elements.

Because of the complicated geometry of an individual tree it is easily understood that the determination of even a vertical profile of LAD is very difficult. With the so-called inversion method (Fritschen 1985), the vertical profiles of wind and radiation are determined as a first step. By assuming a certain vertical profile of these parameters, it is possible to deduce the particular distribution $b(z)$ which results in the best agreement with the observations. Other methods are described by Beadle et al. (1982) and Miller and Lin (1985). There is also the possibility of counting the total number of leaves or needles and surveying the arrangement of branches and trunk (Halldin 1985). All these methods, however, are rather time-consuming and are usually carried out only in exemplary cases.

A further method which takes the stand into account will be introduced below. As outlined in Section 3.1.1, the variables are considered as means over the volume G. In the stand, a portion of G is occupied by branches and leaves. When the volume G_b of these element is known, a volume porosity P may be defined as $P = 1 - G_b/G$. The representative value of a variable ϕ in this approach may be determined from

$$\phi = (1 - P) \cdot \phi_b + P \cdot \phi_0, \tag{3.19}$$

where ϕ_b is the value of a variable in the stand element (with $u_b = v_b = w_b = 0$, $E_b = 0$, $\theta_b = 0$). ϕ_0 may be calculated with the aid of the system of equations given in Section 3.1.1, but without the additional canopy terms in Eqs. (3.9) and (3.13).

For the small trees used in wind tunnel experiments, this porosity may be determined fairly accurately. Ruck and Schmitt (1986a) reported a value of $P = 0.934$. Their method does not furnish any indication on the variation of LAD in space and, consequently, a homogeneous distribution has to be assumed in the numerical simulations. A relation between LAD $b(z)$ and porosity, P may be found by assuming that the two methods described above will supply the same results for stationary horizontally homogeneous conditions. There is:

$$0 = -\frac{1}{\varrho} \frac{\partial p}{\partial x} - \mu u_1 - c_d b u_1^2, \tag{3.20}$$

from which the velocity u_1 may be found as:

$$u_1 = -\frac{\mu}{2 c_d b} + \sqrt{\frac{\mu^2}{4 c_d^2 b^2} - \frac{1}{\varrho} \frac{\partial p}{\partial x} \frac{1}{c_d b}}. \tag{3.21}$$

Table 3.1. Relation between leaf area density b and porosity P (for $C = 1$)

P	0.0	0.1	0.2	0.3	0.4	0.5	0.6	0.7	0.8	0.9	1.0
b	∞	90	20	7.8	3.7	2.0	1.1	0.6	0.3	0.1	0.0

Furthermore, there is:

$$0 = -\frac{1}{\varrho}\frac{\partial p}{\partial x} - \mu u_2. \tag{3.22}$$

As a result of postulating:

$$P \cdot u_2 = u_1, \tag{3.23}$$

LAD is found as:

$$b = C \cdot \frac{1 - P}{P^2}, \tag{3.24}$$

with

$$C = \mu^2 \left(c_d \frac{1}{\varrho}\frac{\partial p}{\partial x} \right)^{-1}. \tag{3.25}$$

For the extreme case of a rigid impermeable body with a zero porosity the above relation results in $b = \infty$, whereas at $P = 1$ (no obstacle) LAD has to approach zero. When typical values are used for c_d, μ and $(1/\varrho)\,\partial p/\partial x$ in Eq. (3.25) (i.e. $c_d = 1$, $\mu = 10^{-2}\,\mathrm{s}^{-1}$, $(1/\varrho)\,\partial p/\partial x = 10^{-4}\,\mathrm{m\,s}^{-2}$), $C = 1$. In Table 3.1 values are compiled for b at certain porosities with $C = 1$.

3.2 Investigations on Single Trees

In a first step, the flow and turbulence distributions around an isolated obstacle are studied. In order to separate the influence of the variation of different input parameters a reference simulation is carried out, with which all other calculations are then compared. The meteorological input parameters and the geometry of the tree may be varied.

In order to obtain an indication of the quality of the simulation, the numerical results are compared, wherever possible, with the data derived by Ruck and Schmitt (1986a) from wind tunnel experiments. This paper will be referred to hereafter as RS-86.

It must be pointed out, however, that such a comparison will be of a qualitative nature only. The numerical simulation results are a volume mean which is compared with a point measurement in the experiment. As a consequence, the observed extreme values are calculated only in an attenuated form. Further-more, it is by no means clear whether the distribution of meteorological variables around a 30-cm-high tree is strictly similar to that around a 15-m-high tree. In addition, the simulations are based on the assumption of constant porosity, whereas the measurements include variations of this parameter. Never-theless, one should expect that the overall structure of the patterns in laboratory experiments and numerical simulations show a very good agreement.

To conform with RS-86 the coordinate system is orientated such that the mean wind is parallel to the x-axis and the wind speed at the upper boundary of the model is taken as 1.2 m s^{-1} for the reference run. The roughness length is set at 0.007 m. As thermal stratification a neutral stratified atmosphere is assumed.

The computation volume is covered by a grid with a unit width of 1 m in both the horizontal and vertical directions. As the x-axis coincides with the wind direction at height H, more grid points are assumed in this direction rather than at right angles to it. The total number of grid points is $81 \times 31 \times 41$ (verti-cal) = 102951. The number is effectively doubled in the y-direction because of the assumed symmetry conditions.

Within the obstacle the individual stand elements cannot be considered as their dimensions are below the cell size of 1 m. It is thus necessary to take an abstract approach, and the flow conditions around porous cone-shaped and ellipsoidal bodies have to be investigated (Fig. 3.3). The shapes are taken as representative for conifers and deciduous trees, respectively. The shape of the obstacle may be selected freely from within fairly wide limits. It is possible to modify the height K and diameter D of the crown, its porosity P and the overall height of the tree. For the reference run a cone-shaped tree with $K = 16 \text{ m}$, $D = 12 \text{ m}$, $P = 0.93$ and $h_s = 0 \text{ m}$ has been used.

As outlined in the preceeding chapter, stationary conditions are simulated, i.e. the forcing by the pressure gradient and ground temperature are kept

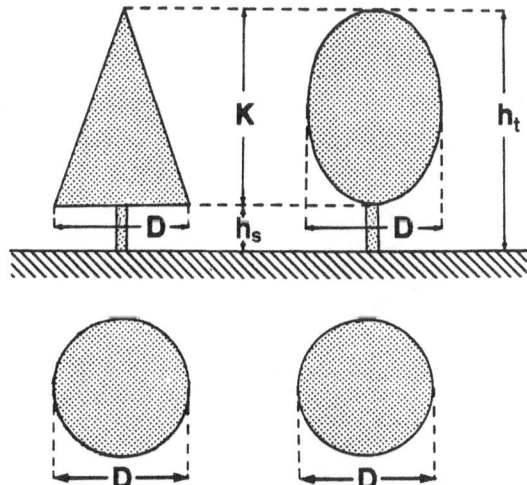

Fig. 3.3. Stand geometry

constant with time. The calculations are terminated after a specific time, allowing an air parcel to move with the vertically averaged velocity from the inflow
boundary through the simulation region four times. Detailed investigations
showed that already by half of this time, any further changes in the simulation
were rather small.

3.2.1 The Reference Run

The horizontal flow field calculated for 1 m above ground level is represented by
the streamlines in Fig. 3.4a. They show only the direction of the wind without

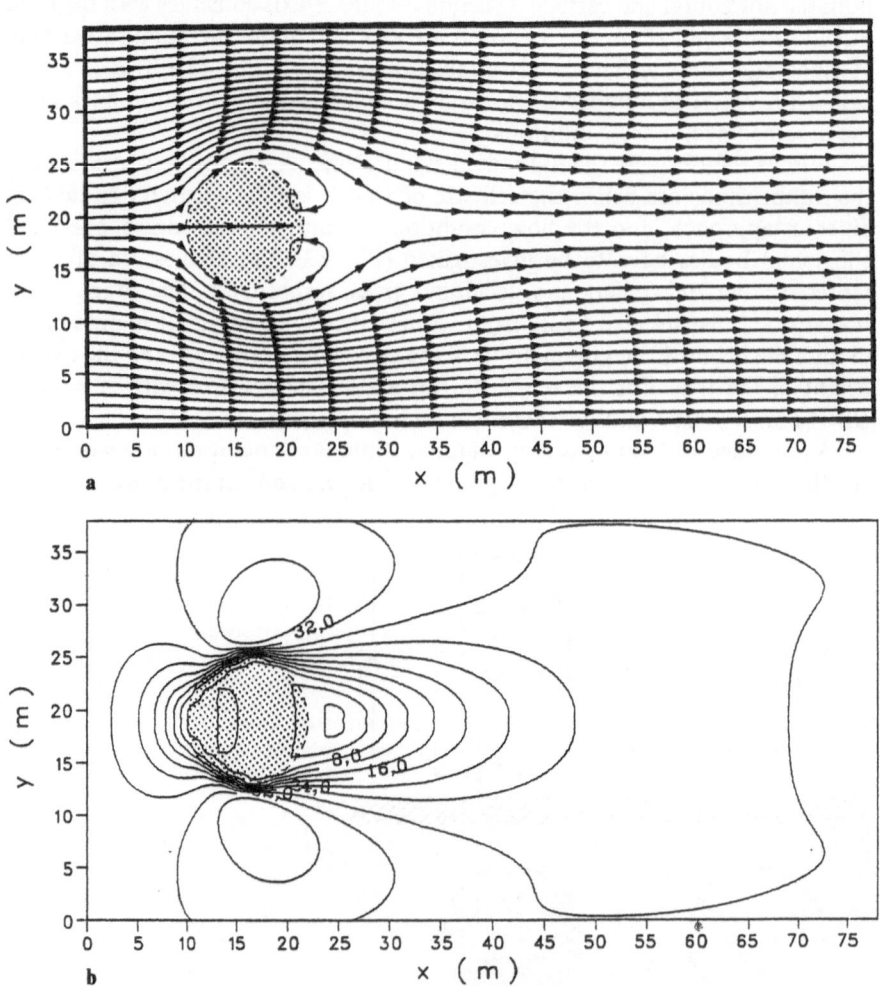

Fig. 3.4 (a) and (b)

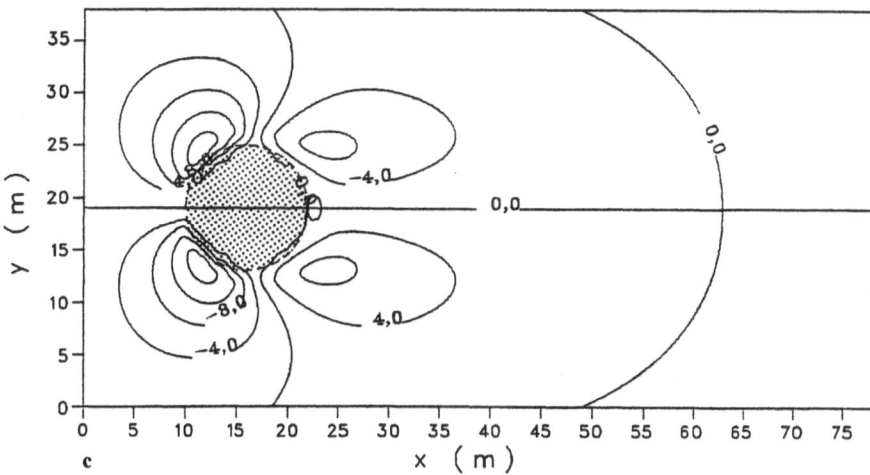

Fig. 3.4. a Streamlines 1 m above ground (cone geometry, neutral stratification) **b** $\bar{u}$ component of velocity 1 m above ground (cm s^{-1}, intervals 4 cm s^{-1}) **c** $\bar{v}$ component of velocity 1 m above ground (cm s^{-1}, intervals 4 cm s^{-1}). *Shaded area* indicates obstacle

indicating the wind speed. The dotted circle in this figure marks the position of the porous model cone. This acts as an obstacle around which the flow mainly takes place; inside the tree only very small velocities are simulated.

The calculated pressure distribution shows a pronounced maximum at the windward side of the cone. This leads to a slowdown of the flow ahead of the obstacle and an acceleration around the tree. At the back third of the obstacle (viewed in the flow direction) the flow becomes detached and an area of reverse flow with a paired eddy is developed.

The extension of the recirculation zone and the distance to which the influence of the obstacle is significant, is shown in Fig. 3.4b and c. The data refer to the $\bar{u}$ and $\bar{v}$ components of the velocity, also at 1 m above ground level. Even at the end of the model area, at $x = 80$ m, small deviations from the undisturbed inflow value are observed.

Additional vertical sections along the symmetry axis provide indications of the three-dimensional distribution of the variables. Based on the simulated field of the $\bar{u}$ component (Fig. 3.5a), the vertical thickness of the reverse flow zone may be delineated. It should be borne in mind that due to the upward decreasing width of the obstacle, the air flow becomes increasingly influenced by the outside air flow. In particular, at the top of the tree, which is represented by only one grid point, only small differences to the inflow value are simulated. The vertical velocity field. (Fig. 3.5b) shows overflowing of the tree with a rise on the windward side and a descent on the leeward side together with the eddy behind the tree.

In this reverse flow zone relatively low velocities are simulated (Fig. 3.6a). Comparing the vertical profiles for $\bar{u}$ at different positions along the symmetry

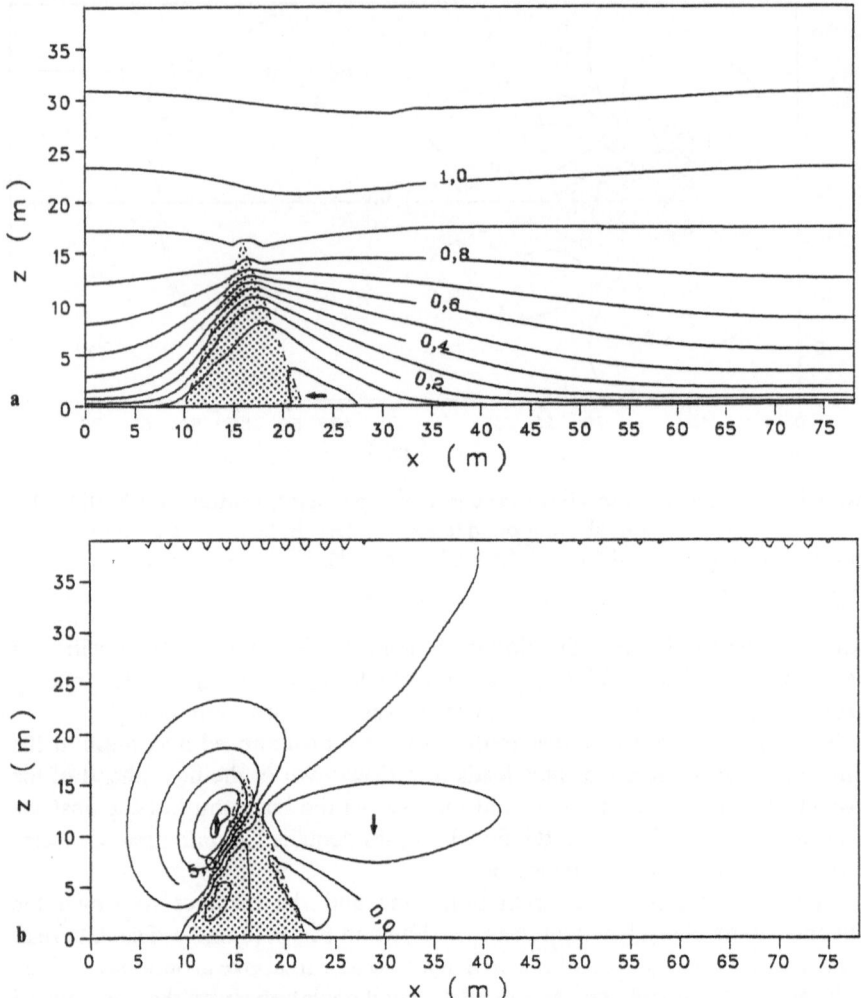

Fig. 3.5. a Vertical section of $\bar{u}$ component of velocity at $y = 19$ m(m s^{-1}, intervals 0.1 m s^{-1}). **b** Vertical section of $\bar{w}$ component of velocity at $y = 19$ m(cm s^{-1}, intervals 2.5 cm s^{-1}). *Shaded area* indicates obstacle

plane with the inflow profile (dashed line), it is found that the velocity defect disappears only gradually in the downstream direction. The large vertical wind shear occurring near the ground in the undisturbed case is shifted upward. As in the equation for the determination of E this parameter describes the transformation of the energy of the mean flow into turbulent kinetic energy, considerably greater turbulence must be expected in this zone. This is clearly evident in the vertical profiles of the turbulent vertical momentum transport (Fig. 3.6b).

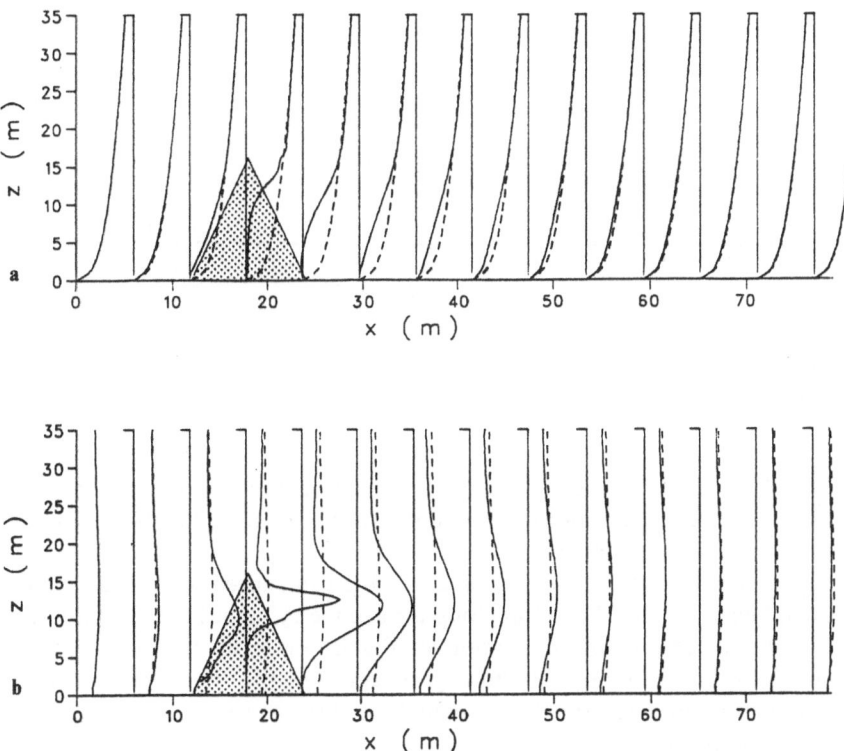

Fig. 3.6. a Vertical profiles of $\bar{u}$ component at $y = 19$ m for cone geometry at neutral stratification (undisturbed initial profile shown by *broken line*). **b** Vertical profiles of shear stress at $y = 19$ m (undisturbed initial profile shown by *broken line*). *Shaded area* indicates obstacle

Measurements of this parameter in wind tunnels (RS-86) and in field experiments (Bradley and Mulhearn 1983; Finnigan and Bradley 1983) show satisfactory agreement with the present simulations (cf. also Chap. 2).

For the same set of input parameters a simulation is carried out in which the stand is not considered in terms of the porosity, but in terms of an additional drag term in Eq. (3.9) and the additional production term in Eq. (3.13). As outlined above, there is the problem that neither the drag coefficient c_d nor the leaf area density $b(z)$ is known. In principle, Eq. (3.24) allows the determination of a leaf area density at a given P. However, when developing these relations, certain assumptions have to be met and these cannot be fulfilled here. Therefore, a number of simulations are performed, where the product $c_d \cdot b$ is modified until a good agreement with the above results is achieved. They may be considered as realistic, because in wind tunnel experiments very similar flow and turbulence patterns are observed.

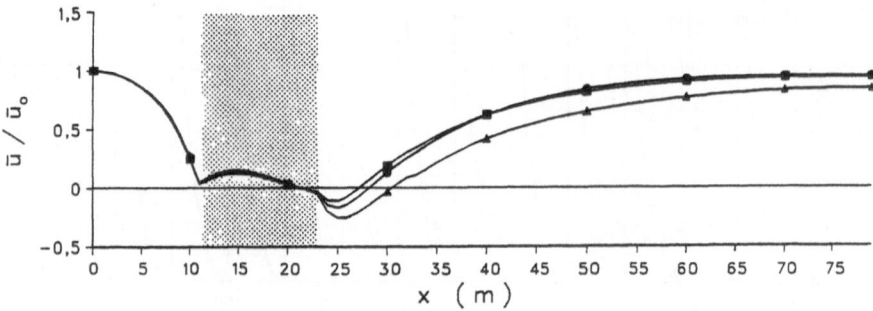

Fig. 3.7. Horizontal profile of $\bar{u}$ component at 1 m height at $y = 19$ m normalized to undisturbed value. *Squares* $c_d \cdot b = 0.1$; *circles* $c_d \cdot b = 0.5$; *triangles* $c_d \cdot b = 1.0$; *broken line* for $P = 0.93$). *Shaded area* indicates the width of the obstacle

For $P = 1$, $c_d \cdot b$ is varied between 0.1 and 2. Evaluation of the results shows that when increasing the product to above 1, only minor differences from the case $c_d \cdot b = 1$ are calculated.

Figure 3.7 illustrates the development of the wind speed (normalized to the inflow velocity) at a height of 1 m along the symmetry plane for different values of $c_d \cdot b$. The corresponding simulation for $P = 0.93$ is given by the dashed line. Although the horizontal profile shows good agreement for small values of $c_d \cdot b$, this simulation does not allow the calculation of the downwind eddy for small values of the product.

When $c_d \cdot b$ is increased from 0.5 to 1.0, the agreement improves markedly. This also applies to three-dimensional distribution of the meteorological variables.

When using a typical vertical profile of $b(z)$, as used in Rauner (1976) or Grin et al. (1970), a characteristic mean leaf area density of 1 $m^2 m^{-3}$ is found at a leaf area index of 10. This implies that in the present simulation the drag coefficient c_d must be of the same order of magnitude.

The drag coefficients for different types of trees may, in principle, be determined in wind tunnel experiments (Ylinen 1952; Panggabean 1978). Here, the wind force F exerted on the isolated obstacle is usually measured and then related to the dynamic pressure. As Mayhead (1973) pointed out, the difficulties in determining c_d become quite obvious in this case. The definition and exact determination of characteristic values for mean wind speed, wind force and leaf area are rather difficult. This results in the considerable range of values of c_d found in the literature. There is, however, a general trend for a decrease of c_d with increasing wind speed. This can be ascribed to the tendency of leaves and branches to become oriented in the direction of the wind with increasing mean velocity, leading to a reduction of the effective cross section of the leaf area.

Drag coefficients obtained from observations are in the range 0.2–2.0 (Thom 1968, 1971; Meroney 1968; Kondo and Akashi 1976; Seginer et al. 1976). As outlined above, smaller values of c_d correspond to higher wind speeds and vice versa. For the low values for the superimposed wind speeds used here, a drag coefficient of 1 appears realistic.

The results described above show that the air flow is around and above the tree, which acts as a large obstacle. Even inside the porous body low wind speeds are simulated and this means that part of the oncoming wind goes through the tree. In the case, where the lower part of the crown is elevated above the surface by the trunk, flow can also take place below the dense crown.

From the calculated velocity field it is possible to extract quantitatively the percentage of flow which takes place around, over, through or under the tree. For this purpose a vertical section is placed through the tree center at right angles to the large-scale flow in which the following areas are then defined (Fig. 3.8): F_1, flow over the obstacle, F_2, flow around the obstacle, F_3, flow through the obstacle and F_4, flow under the obstacle.

Each grid point of this section is assigned to one of these areas and the corresponding calculated $\bar{u}$ components are cumulated. At the same corresponding grid points on the inflow boundary also sums are performed, indicated by the additional index o, and these results are then compared (Table 3.2).

The relatively large value of 0.18 for flow through the tree can be ascribed to the fact that near the top of the tree the outer flow penetrates to the tree's interior where the wind suffers only little further retardation. At the widest extension of the crown, the wind speed is only about 1–5% of that at the corresponding height along the inflow boundary.

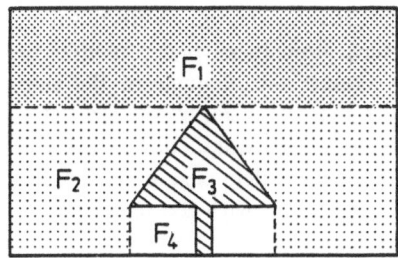

Fig. 3.8. Schematic definition of areas used in calculation of flow above, around, through and below an isolated tree

Table 3.2. Mean mass flux around, over and through a trunkless porous cone under neutral stratification relative to the undisturbed values at the inflow boundary

F_1/F_{1o}	F_2/F_{2o}	F_3/F_{3o}	F_4/F_{4o}
1.02	1.05	0.18	0.00

3.2.2 *Variation of Meteorological Input Parameters*

Among the meteorological input parameters, the larger-scale wind speed and
the thermal stratification of the basic state may be varied. In wind tunnel
investigations the influence of stratification on flow and turbulence fields can be
measured only with great difficulty. In numerical simulations, however, temperature stratification can be taken into account fairly easily. As a measure for the
thermal stratification the vertical gradient of the potential temperature $\partial\tilde{\theta}/\partial z$ is
used. It is positive for stable stratification and negative for unstable conditions.

As a consequence of the changed temperature stratification a modified
logarithmic wind profile is obtained. Under unstable conditions strong vertical
mixing (characterized by a large diffusion coefficient) leads to higher wind
velocities close to the ground, whereas with increasing stability the profile of
$\bar{u}(z)$ tends more to a linear pattern.

Thermal stratification has a pronounced influence on the degree to which
flow takes place around or over the obstacle. As shown by Gill (1982), a certain
kinetic energy can be assigned to a volume of air. When the kinetic energy of an
air parcel is sufficient to surmount the height of the obstacle, a corresponding
portion is transformed into potential energy. If the energy is not sufficient, the
flow is blocked and an eddy forms on the windward side. In the case of
$\partial\tilde{\theta}/\partial z > 0$, additional work has to be performed against the stratification. At the
same time, the kinetic energy near the ground is smaller than for neutral
conditions and flow will take place around rather than over the obstacle.

The influence of changes in velocity on the wind field around an isolated tree
can be deduced from laboratory experiments. Homann (cited in Prandtl 1965)
investigated the flow behind a cylinder at different velocities (Reynolds numbers,
Re). At small *Re* numbers (i.e., low velocities), flow takes place around the
cylinder as potential flow which closes up against the obstacle in the same
manner as it has opened up in front of it. With increasing *Re* number, a paired
eddy is established on the leeward side. The associated wake zone grows with
increasing velocity until eventually closed eddies are detached in alternating
fashion from either side of the cylinder and drift along with the mean flow in
a Karman vortex street. Regular vortex streets, however, are only possible for *Re*
numbers between 60 and 5000.

With all other input parameters remaining as in the reference run, the
superimposed wind speed $\bar{u}(H)$ was reduced to 50% in one simulation and
doubled in the other. The influence of these altered wind velocities on the
calculated fields of the meteorological variables will be discussed with particular
regard to the reverse flow area behind the cone.

In Fig. 3.9 the horizontal profiles of the $\bar{u}$ component of the wind are
presented for 1 m above ground along the symmetry axis. The values are
normalized to the respective values at the inflow boundary. The profiles are
rather similar for all three velocities: An area with strong retardation on the
windward side of the obstacle is followed by a zone with very low velocities
within the obstacle itself. Larger differences are developed only behind the

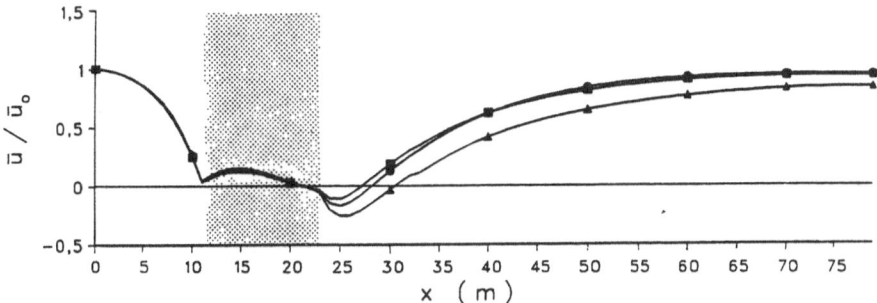

Fig. 3.9. Horizontal profile of $\bar{u}$ component at 1 m height at $y = 19$ m, normalized to undisturbed value (cone-shaped tree, neutral stratification). *Squares* $\bar{u}(H) = 0.6\,\mathrm{m\,s^{-1}}$; *circles* $\bar{u}(H) = 1.2\,\mathrm{m\,s^{-1}}$; *triangles* $\bar{u}(H) = 2.4\,\mathrm{m\,s^{-1}}$. *Shaded area* indicates the width of the obstacle

obstacle, the zone with $\bar{u} < 0$ growing in size with the increase in the wind speed selected. The simulations thus result in situations similar to those observed in laboratory investigations and referred to above. With the aid of the diffusion coefficient, a *Re* number, $Re = \bar{u}(H) \cdot D/K_m$, characteristic for atmospheric flows may be defined (Atkinson 1981). In the numerical simulations presented here, this is in the range 10–20 and thus no vortex streets are established.

The extent of the simulation region on the downwind side of the cone corresponds to about four times the tree diameter ($x/D = 4$). Even at $x = 80$ m, velocity differences to the initial value on the inflow side may be noted.

However, similar effects have been observed in the laboratory and in the field. In wind tunnel experiments, Meroney (1968) found that for flow over an 18-cm-high model tree the undisturbed conditions had not been reestablished even after $x/D = 10$. Studying the flow conditions around a pine branch at wind speeds between 1.4 and 4.5 m s^{-1}, Grant (1983) observed disturbances even at a distance of $x/D = 18$. In field experiments with a porous fence, Bradley and Mulhearn (1983) found that even at a distance equivalent to 50 times the fence height the flow had not yet returned to the undisturbed original situation. The selection of the wind velocity controls not only the horizontal extent but also the vertical thickness of the reverse flow zone on the downwind side. When increasing $\bar{u}(H)$ from 0.6 to 2.4 m s^{-1} the leeside recirculation zone doubles in thickness (Fig. 3.10).

A number of characteristic values of the leeward eddies at different wind speeds are summarized in Table 3.3. One finds a near-linear relation between the horizontal extent X of the reverse flow zone and the larger-scale velocity. This also applies to $\bar{u}_{\mathrm{min}}/\bar{u}(H)$.

In order to study the influence of thermal stratification on the distribution of the variables, the vertical gradient of the potential temperature of the basic state is increased from $0\,\mathrm{K}/100$ m to $1\,\mathrm{K}/100$ m. Considering that the standard atmosphere is based on a gradient of 0.35 K/100 m, the value selected here

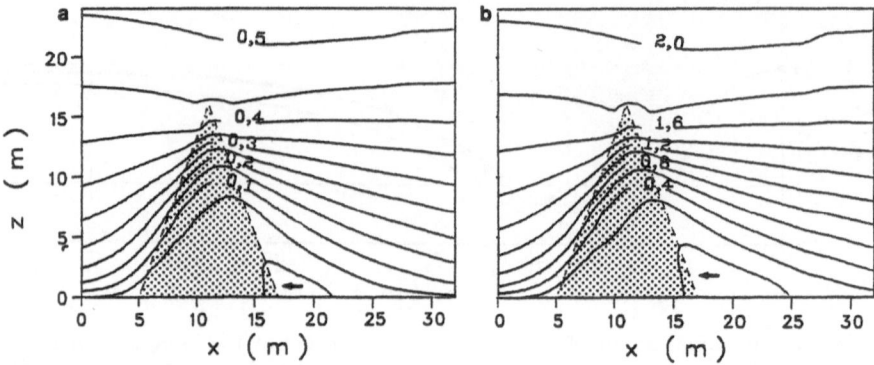

Fig. 3.10. Vertical section of $\bar{u}$ component of velocity (m s^{-1}) at $y = 19$ m for cone-shaped tree under neutral stratification. **a** For $\bar{u}(H) = 0.6$ m s^{-1} (intervals 0.05 m s^{-1}); **b** for $\bar{u}(H) = 2.4$ m s^{-1} (intervals 0.2 m s^{-1})

Table 3.3. Minimum wind speed and largest extent X of reverse flow area for different values of $\bar{u}(H)$ (cone-shaped tree, neutral stratification)

$u(H)$	m s^{-1}	0.6	1.2	2.4
u_{min}	cm s^{-1}	-1.5	-4.8	-16.5
$u_{min}/u(H)$		2.5×10^{-2}	4.0×10^{-2}	6.9×10^{-2}
X	m	5.4	6.5	8.6

represents a rather stable stratified atmosphere. The value is, however, by no means unrealistic, as at night much larger gradients are developed near the ground in unforested areas.

As outlined above, in a stable stratified atmosphere reduced turbulent mixing leads to wind speeds, which, near the ground are generally lower than in the neutral case. Figure 3.11 shows that the velocity is nearly halved at 1 m above ground level. As a consequence of this lower undisturbed wind speed, a pronounced recirculation zone is simulated on the leeward side of the obstacle (Fig. 3.12). Vertical movements are suppressed because for upward deflections work has to be done against the thermal stratification. The flow will thus tend to circumvent the obstacle horizontally to a larger degree. Figure 3.13 illustrates the horizontal mass flux profile through a vertical plane oriented perpendicular to the superimposed wind at $x = 16$ m. It compares the results for simulations with neutral and stable stratification. It is clearly evident that for the case of the isolated tree, flow in the stable case occurs to a greater extent around the obstacle than it does in the calculations with $\partial\bar{\theta}/\partial z = 0$. It becomes obvious from Eq. (3.13) that turbulence decreases with increasing stability. A comparison of the vertical profiles of the Reynolds stress for the two simulations with different thermal stratifications is shown in Fig. 3.14. While the structures of the

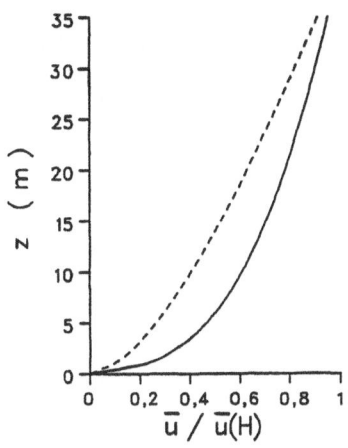

Fig. 3.11. Vertical profiles of $\bar{u}$ component of velocity for different conditions of thermal stratification. *Dashed line* Stable; *solid line* neutral stratification

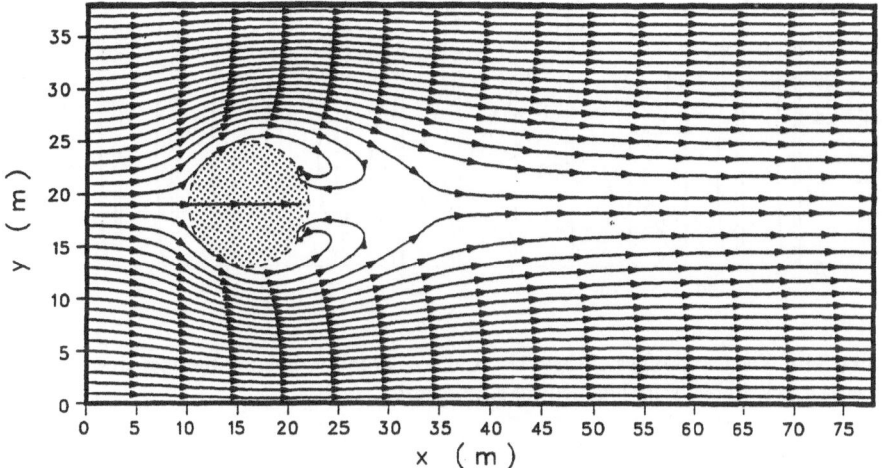

Fig. 3.12. Streamlines 1 m above ground for flow around a cone-shaded tree (*shaded area*) under strongly stable stratification

corresponding profiles agree rather well, especially as far as the maxima are concerned, the calculated values in the neutral case are about twice those of the stable case.

3.2.3 Variations in Shape of the Individual Tree

The results of the preceeding paragraph show that differences in the meteorological input parameters influence the velocity distribution significantly. It may

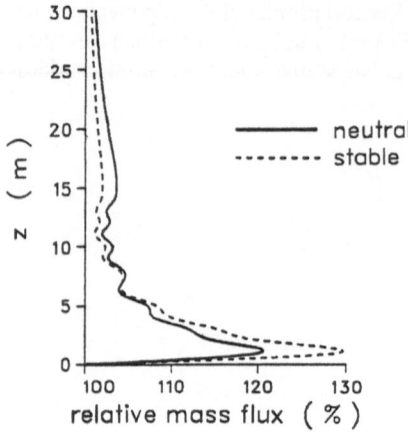

Fig. 3.13. Vertical profiles of relative mass flux through vertical plane F_2 (cf. Fig. 3.8) at $x = 16$ m for different thermal stratification

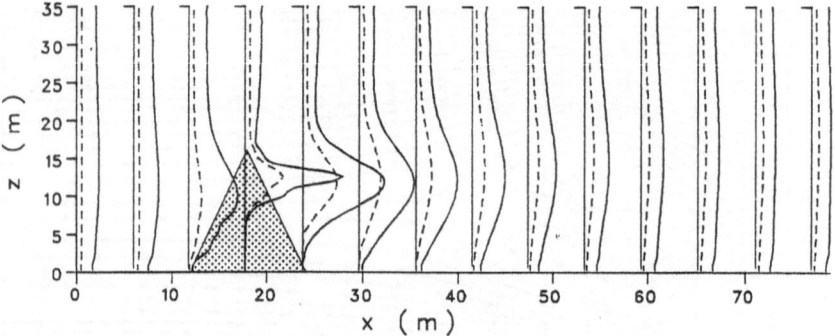

Fig. 3.14. Vertical profiles of shear stress at $y = 19$ m for different thermal stratification. *Dashed line* Stable; *solid line* neutral stratification

be stated, however, that for the three-dimensional distribution, the variables show similar distribution patterns and differ only in their extremes.

However, when the shape of the tree is changed, much larger modifications have to be expected. This is due to the fact that a taller tree (crown diameter, tree height) represents a more pronounced obstacle to the flow. Larger disturbances also have to be anticipated with a decrease in porosity.

The tree shape used so far in the simulations is only rarely encountered in nature. It was chosen, because it was rather similar to the obstacle in the wind tunnel experiments by RS-86. In most natural cases, leaves and branches only start from a certain height above the ground, and below this a trunk is the only obstacle to the flow. This implies that an air flow is possible not only over and around the tree, but also under it. In the trunk zone, the reduced resistance frequently leads to higher wind speeds than in the crown above. The respective measurements for individual trees are found in RS-86; other authors have also observed the same phenomenon in groups of trees and in forests (Fons 1940; Reifsnyder 1950; Geiger 1961; Oliver 1971).

This jet flow in the trunk zone may also be simulated with a trunk height of, e.g., 6 m. All other input parameters remain as in the reference run. In Fig. 3.15 the flow is represented by wind vectors (indicating the horizontal ($\bar{u}$) and vertical ($\bar{w}$) wind components) in a vertical x–z section along the symmetry plane. In addition to the velocity maximum in the trunk zone, there is a zone of reverse flow elevated from the ground. The velocity maxima below the crown are almost double the values found at the same height in the undisturbed profile. This large value is caused by the strong downward transport of momentum ahead of the tree. This result is also in excellent agreement with the laboratory data of RS-86. These authors report that when underflow is allowed, the disturbance of the atmospheric boundary layer profile is more pronounced than in the case without a trunk.

This observation is also supported by the simulations. In Fig. 3.16 horizontal profiles of the $\bar{u}$ component of the velocity are shown for different trunk heights (0, 3 and 6 m). The disturbances in the immediate vicinity of the porous cone are largest for a vanishing trunk height. The profile at $x = 24$ m shows very low velocities close to the tree for $h_s = 0$ m, as here the reverse flow zone is fully developed. However, in the profiles for $h_s = 3$ m and $h_s = 6$ m, the wind shadow effect of the trunk is clearly recognizable. At a distance of about one crown diameter, the velocity deficit may be approximated by a Gaussian distribution, regardless of h_s. Meroney (1968) observed that in laboratory experiments, this occurs at a distance of 3–4D. For shorter distances he found a skewed distribu-

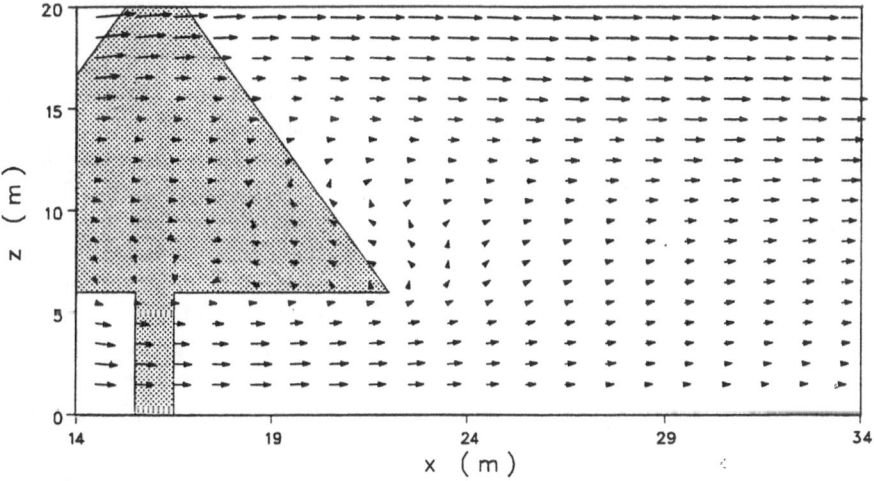

Fig. 3.15. Wind vectors ($\bar{u}$ and $\bar{w}$ components) in a vertical section (only a subsection of the total model domain is shown) at $y = 19$ m around a cone-shaped tree under neutral stratification

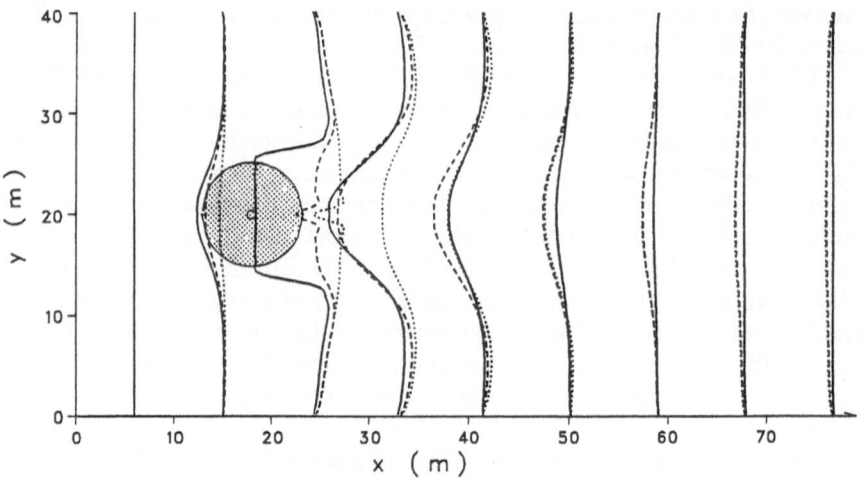

Fig. 3.16. Horizontal profiles of $\bar{u}$ component of velocity for different trunk heights. *Solid line* 0 m; *dashed line* 3 m; *dotted line* 6 m

tion; however, this cannot be calculated here due to the symmetry assumption used in the simulations.

The zone of wind shadow widens in a roughly linear fashion in the direction of the mean flow, while, at the same time, the velocity deficits diminish in size. Whereas at $h_s = 0$ m the disturbances in the immediate vicinity of the tree were larger than for $h_s > 0$ m, this relation is reversed beyond a distance of about $2D$. At the outflow border of the integration region, the velocity disturbances are larger than in the case without a trunk due to the underflow allowed in this simulation.

Outside the tree, turbulence is generated under neutral stratification only through transformation of the kinetic energy of the mean flow. Because of the use of the flux-gradient relationship, the vertical turbulent momentum flux is largest where the mean wind shear attains its maximum. In the boundary layer close to the ground, one usually observes a downward momentum flux. This is also true throughout the integration region in the simulations carried out for a trunkless tree. When underflow is allowed, the situation becomes more complicated due to the resulting jet in the trunk zone. Here, in certain regions, an upward turbulent momentum flux is also possible (Fig. 3.17).

When the profiles for a certain distance are compared, good qualitative agreement is found when the profile in the trunkless case is shifted upward by h_s. As in RS-86, larger values are simulated for $h_s > 0$ than for $h_s = 0$ m. The authors explain this by the increased formation of eddies when a trunk is present. In constrast to this, during flow predominantly around the obstacle an attenuation of the eddy advance takes place mainly along the ground.

The influence of the variation of crown diameter and height on the flow field for neutral stratification may be summarized as follows: In all simulations in the

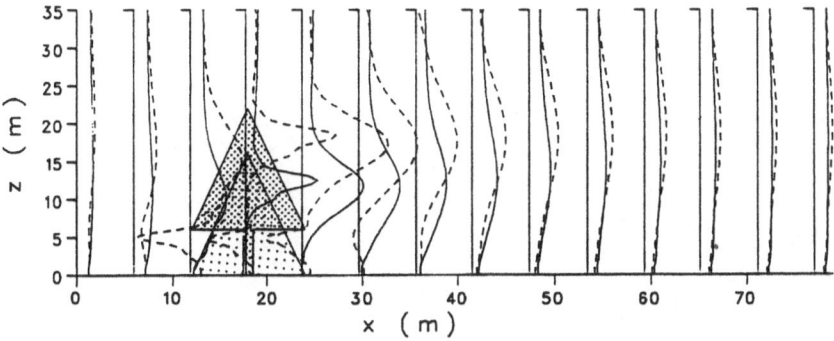

Fig. 3.17. Vertical profiles of shear stress at $y = 19$ m. *Solid line* With trunk; *dashed line* without trunk

Table 3.4. Maximum extent (X, Y, Z) of the leeward zone of reverse flow and minimum velocities for different crown shapes (neutral stratification)

K (m)	D (m)	u_{min} (cm s^{-1})	X (m)	Y (m)	Z (m)
16	13	− 4.8	7	7	3
10	13	− 2.4	4	7	2
20	13	− 6.8	9	7	4
16	8	− 2.9	4	3	2
16	13	− 5.8	9	11	5

downwind direction a reverse flow is found, the extent of which is shown in Table 3.4.

According to this, the width of the reverse flow zone is controlled mainly by the crown diameter, whereas its length and vertical extent depend to the same degree on the dimensions of K and D. The larger the obstacle, the more pronounced will be the deviations of the calculated variables at the outflow boundary from the undisturbed conditions. The largest differences of the $\bar{u}$ component 1 m above ground level at $x = 80$ m in the reference run are in the range of 5%. For smaller trees they are reduced to 1–3%, while they increase up to 12% for larger trees.

As an additional parameter, the porosity P of the individual tree is changed and this is a measure of the density of the foliage. The most pronounced change is calculated when P is reduced from 1.0 to 0.9. A porosity different from 1.0 implies that in a horizontally homogeneous region an obstacle of a certain shape is established. The velocity will be reduced within the obstacle and at a certain porosity the values will become so low that any further reduction of P will no

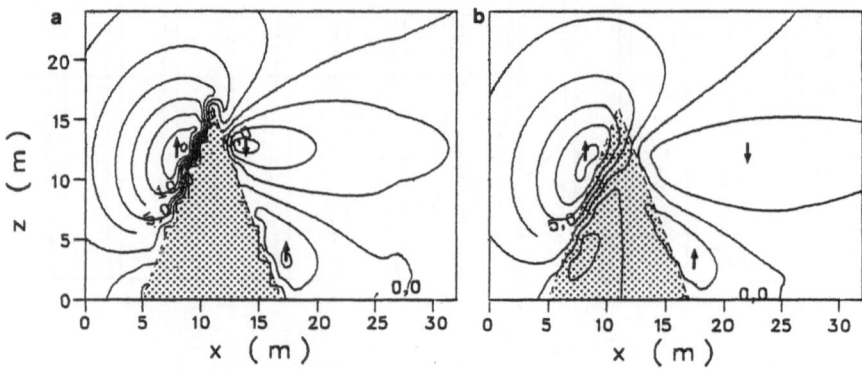

Fig. 3.18. Vertical section of the $\bar{w}$ component (cm s^{-1}, intervals 2.5 cm s^{-1}) at $y = 19$ m for an isolated cone-shaped tree under neutral stratification. With **a** $P = 0$; **b** $P = 0.93$

longer lead to significant changes. For the limiting case of a solid body ($P = 0$) flow can no longer go through the obstacle but occurs entirely around it. Correspondingly larger values are achieved by the $\bar{v}$ component of the velocity, the maximum increasing from 0.2 m s^{-1} at $P = 0.93$ to 0.26 m s^{-1} at $P = 0.0$. The extreme values of the vertical velocity $\bar{w}$ increase accordingly. In both simulations large differences are found especially for the upper part of the crown. At $P = 0.93$ the outside air flow intrudes into the foliage where it is only retarded a little as the uppermost part of the crown represents only a small obstacle. Naturally, the situation is different for $P = 0$ when the wind in the interior of the tree is zero. As a consequence, the horizontal divergence on the leeward side must be compensated, mainly by vertical advection (Fig. 3.18).

In the simulations discussed so far, the obstacle selected was similar in shape to a conifer. In the following the shape will be modified so that the results represent the conditions for a deciduous tree. For this purpose the cone will be substituted by a porous ellipsoid. At identical diameter and crown height, this obstacle represents a larger volume and increased effects on wind and turbulence must expected.

All meteorological input parameters are the same as for the reference run. Also, characteristic values for the tree ($P = 0.93$, $K = 16$ m, $D = 13$ m, $h_s = 0$ m) remain the same. However, now the largest diameter of the crown is not at the ground but at a height $K/2$.

The air flow close to the ground, illustrated by the streamlines 1 m above ground level, shows some differences from the corresponding simulations for the cone (Fig. 3.19). Because of the influence of the geometry of the individual tree on the flow field discussed above, it is not surprising to learn that the leeward area of reverse flow is now longer but not so wide. This is due to the fact that near the ground the ellipsoid is narrower than the cone, but on the whole it represents a larger obstacle.

Simulations with a trunk ($h_s = 3$ m) shows a pronounced velocity maximum in the trunk zone and, as for the cone, an elevated reverse flow area with lower

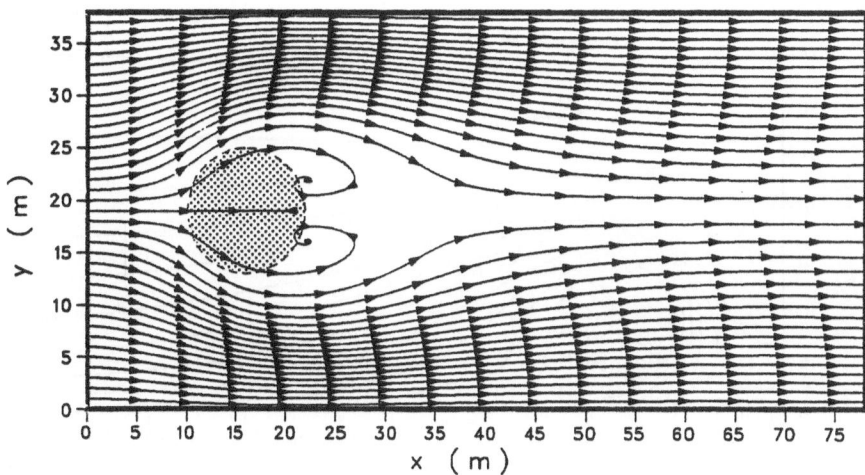

Fig. 3.19. Streamlines 1 m above ground (ellipsoidal tree without trunk, neutral stratification)

velocities (Fig. 3.20a). The vertical cross section for the $\bar{w}$ component (Fig. 3.20b) shows that the flow on the windward side is split in half almost exactly at half the crown height, one part moving under the obstacle and the other part moving over it. A deviation to the left or right, seen in the flow direction, is not possible in this vertical section as the $\bar{v}$ component disappears along the symmetry plane.

In Fig. 3.21 the wind profiles for both types of tree are compared. It again becomes obvious that the geometry of the crown has a decisive influence on the flow pattern. The velocity deficit behind the more voluminous ellipsoidal geometry is larger than behind the cone. For the latter, the effective cross-sectional area decreases linearly with height. Even along the outflow boundary this effect still appears.

In an analogy to the cone, the geometric parameters of the crown were also modified for the ellipsoid. However, the results do not offer any new insights, but only support the conclusions drawn for the flow around and under the cone-shaped obstacle.

The effect of a modification of the porosity P within the obstacle on the overall structure of the meteorological fields was also investigated. For this purpose, the ellipsoidal tree was hollowed out. It was assumed that the foliage was concentrated in the marginal areas with $P = 0.93$, while the interior was taken as being free of stand elements.

The simulated velocity field is almost identical to that found for the homogeneous crown. The differences for the whole area investigated, with the exception of the crown, are considerably less than 1%. It may therefore be concluded that once the porosity has fallen below a value of 0.97–0.95, its influence on the calculated distribution of the meteorological variables may be neglected.

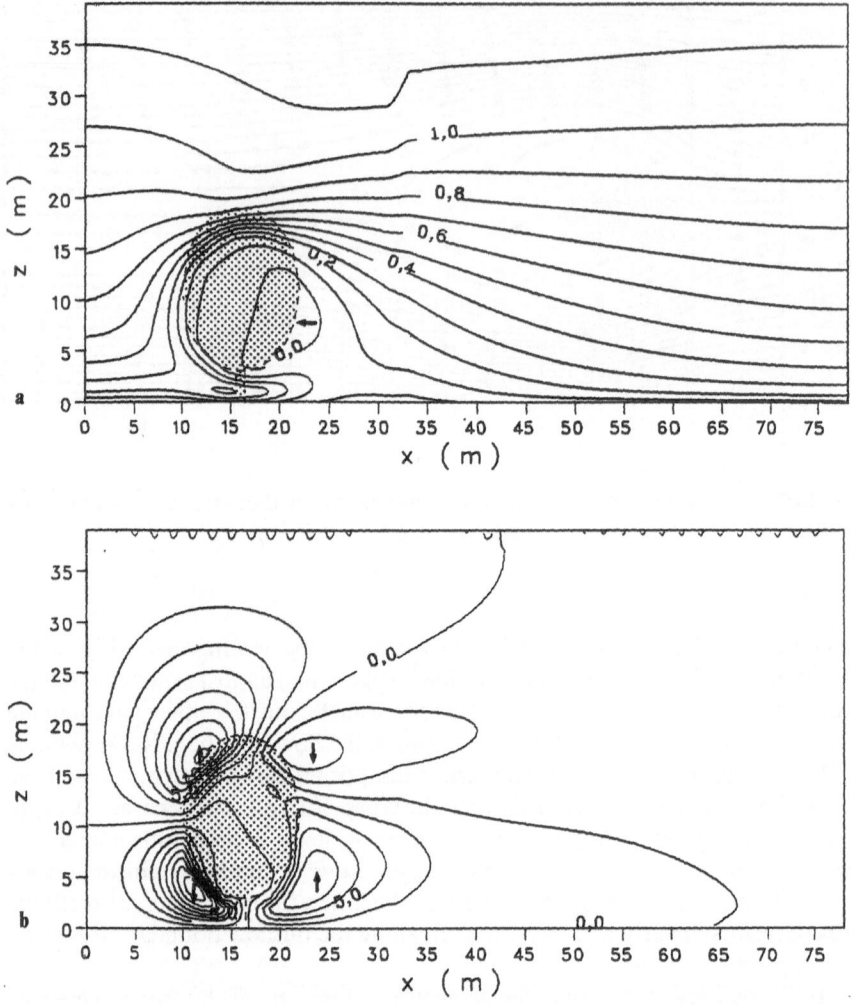

Fig. 3.20. Vertical section of velocity components at $y = 19$ m (ellipsoidal tree, neutral stratification). **a** $\bar{u}$ component (m s^{-1}, intervals 0.1 m s^{-1}); **b** $\bar{w}$ component (cm s^{-1}, intervals 2.5 cm s^{-1})

Furthermore, this implies that the geometry of the tree (K, D, h_s) plays a more dominant role than the porosity of leaf area density which çan be determined in nature only with great difficulty. However, up to now, this statement holds only for flow around an isolated tree.

Even when the flow field outside the tree is changed to a minor degree only, there are still differences within the leaf-free zone in the stand itself. Due to the absence of additional drag, the air flow here is not retarded so much and an additional velocity maximum is simulated (Fig. 3.22).

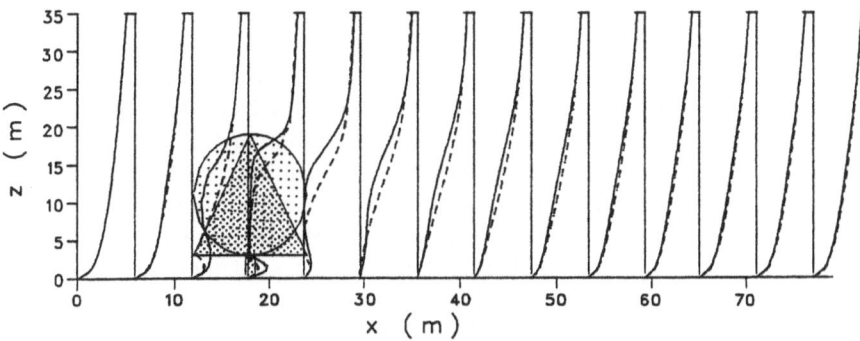

Fig. 3.21. Vertical profiles of $\bar{u}$ component at $y = 19$ m. *Solid line* Deciduous tree; *dashed line* conifer

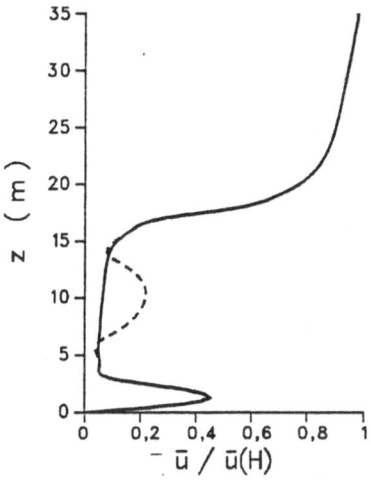

Fig. 3.22. Vertical profiles of $\bar{u}$ component normalized to $\bar{u}(H)$ at $x = 16$ m and $y = 19$ m, neutral stratification. *Solid line* Dense crown; *dashed line* hollow crown

3.2.4 Dispersion of Air Pollutants Around an Isolated Tree

The transport of air pollutants in the atmosphere is controlled by mean wind speed and direction, turbulence conditions and deposition velocity, when the possibility of chemical reactions between compounds is not taken into account. The smaller the aerosol particles, the more they follow the air flow. A reduction in velocity leads to an increase in sedimentation.

When such passive chemical air admixtures come into the area of influence of an isolated tree, increased deposition will take place in zones of reduced wind speed. A stand thus exerts certain filtering influences, as reported by Neuberger et al. (1967), Herbst (1965) and Keller (1970). This applies in particular to the exposed positions along forest margins and near the top of a canopy; this effect diminishes almost exponentially toward the interior of the stand (Baumgartner 1973). The inner part of the stand, which is thus sheltered from emissions, will therefore possess less polluted air than the open ground surrounding it.

In strongly polluted areas, especially around industrial centers, this filtering property can have catastrophic effects for the stand. In the worst case, the excessive deposition of pollutants on the stand elements can lead to the death of the tree. Inventories of forest decline (BML 1984; Schöpfer and Hradetzky 1984) give clear indications for a connection between the shape of the forest edge (crown roughness, stepped profile) and the degree of damage to individual plants. Where such a relation can be observed, correspondingly adjusted plantations may minimize the damage. This conclusion is supported by the wind tunnel experiments of Ruck and Schmitt (1986b), henceforth referred to as RS-86b, in which deposition around isolated trees and in stands with different tree configurations was investigated.

Laboratory experiments on the dispersion of pollutants within and around stands were described by Meroney (1968). Field measurements were carried out by Allison et al. (1968) in a dense rainforest and in deciduous and coniferous forests by Neuberger et al. (1967) and Raynor et al. (1970, 1974).

For the study of the dispersion of air pollutants in forested areas, a Lagrangian dispersion model is combined with the actual flow model. The total mass of a certain pollutant is represented by a number of particles, the trajectories of which may be calculated. When the three-dimensional wind and turbulence patterns are known it is possible to calculate these trajectories. By counting the number of particles present at a certain time in a given volume, the concentration may be simulated. On larger scales such a coupling between a dispersal model and a complicated wind model has been carried out successfully (e.g., Yamada 1985; Gross et al. 1987).

Only a short description of the dispersion model will be given here. For a more in-depth treatment reference should made to the extensive literature on this subject (e.g., Legg and Raupach 1982; Legg 1983; Glaab 1986).

The coordinates x_i of a particle at a time $t + \Delta t$ may be calculated from the position at time t according to

$$x_i(t + \Delta t) = x_i(t) + U_i(t)\Delta t. \tag{3.26}$$

In this equation U_i is the velocity of the particle and Δt the time step.

The particle velocity is split into a mean portion $\bar{u}_i$ (averaged in space and time) and a turbulent deviation u'_i. While $\bar{u}_i$ is supplied directly by the wind model, the turbulent deviation is determined by

$$u'_i(t + \Delta t) = R_{Li}(\Delta t)u'_i(t) + \sqrt{1 - R_{Li}(\Delta t)^2}\,\sigma_{ui}\Omega + [1 - R_{Li}(\Delta t)]T_{Li}\frac{\partial \sigma_{ui}^2}{\partial x_i}, \tag{3.27}$$

where Ω is a Gaussian distributed random number supplied by a random number generator.

The Lagrangian autocorrelation function R_{Li} is assumed to be of an exponential nature according to

$$R_{Li}(\Delta t) = \exp(-\Delta t/T_{Li}). \tag{3.28}$$

The time scale T_{Li} is obtained from the combination of the Taylor theorem with the regularities of Fickian diffusion

$$T_{Li} = K_m \sigma_{ui}^{-2}. \tag{3.29}$$

Besides the mean velocity, the turbulent diffusion coefficient K_m is also supplied by the wind model.

For the determination of the velocity variances σ_{ui}, an empirical approach (e.g., Caughey et al. 1979; Hanna 1981; Vogel 1986) or a relation according to Mellor and Yamada (1982) may be used.

The latter authors theoretically developed a formula which is applicable only in a horizontally homogeneous area. However, it is used here, because in contrast to other approaches, it permits the inclusion of local conditions. The various components are

$$\sigma_u^2 = \frac{E}{6} + \frac{1}{\sqrt{0.5E}}(4P_u - 2P_v + 2P_\theta),$$

$$\sigma_v^2 = \frac{E}{6} + \frac{1}{\sqrt{0.5E}}(-2P_u + 4P_v + 2P_\theta),$$

$$\sigma_w^2 = \frac{E}{6} + \frac{1}{\sqrt{0.5E}}(-2P_u - 2P_v - 4P_\theta), \tag{3.30}$$

with P_u, P_v and P_θ being abbreviations for:

$$P_u = K_m \left(\frac{\partial \bar{u}}{\partial z}\right)^2,$$

$$P_v = K_m \left(\frac{\partial \bar{v}}{\partial z}\right)^2, \tag{3.31}$$

$$P_\theta = \frac{g}{\theta} K_h \frac{\partial \theta}{\partial z}.$$

The last term in Eq. (3.27), which disappears under homogeneous turbulence conditions, covers the influence of velocity variances in space. Neglecting it could lead to an enrichment of particles in zones of reduced turbulence.

The filtering effect of a forest results from the fact that aerosols, dust and gases are deposited on branches and leaves, thereby being withdrawn from further dispersion (Mitscherlich 1975). The efficiency of this sedimentation is controlled by particle size and wind speed. A detailed compilation of wind tunnel experiments and field observations on the deposition of solid, liquid and gaseous air admixtures was prepared by Chamberlain (1975). He also mentions laboratory experiments in which the deposition of particles on plant elements was investigated at different wind speeds. According to this, deposition is greater at lower velocities than at higher ones, because leaves are oriented more into the mean wind direction in the latter case. This reduces the potential impact area for the particles, thereby decreasing the probability of deposition. Similar results were observed by other authors such as Schuepp (1982).

Because of the vast number of possible combinations of particles (type, size, deposition velocity), stand (type, shape, density of foliage) and meteorological parameters (wind speed, turbulence, temperature) it is not possible to obtain exact data on the ensuing sedimentation. Thus, in the dispersion calculation two extreme cases are calculated. Firstly, it is assumed that at the stand margins the air admixture becomes absorbed entirely and that the interior of the tree is free of particles. In the second case, deposition is not permitted and all particles take part in further dispersion.

As a constraint along the boundaries of the integration volume it was assumed that the lateral boundaries are open, i.e., particles may leave the volume, while the ground and the top boundary are impermeable, i.e., the particles are reflected from it. The simulation is stopped when all started particles have left the model domain.

The canopy model presented earlier supplies the required meteorological input parameters such as wind, temperature and turbulence, on a grid extending into all three dimensions and with a unit interval of 1 m. This grid is also used in the dispersion calculation for a point source at height h_q on the windward side of the tree. The concentration may be found by counting the particles within each box of the grid. The input data required for these calculations are summarized in Table 3.5.

The strength of the source is not required as the calculated concentrations are given as a percentage of the maximum concentration. As the largest concentrations are calculated in all simulations with the same amount near the source, the resulting data are directly comparable. It should be remembered, however, that the absolute amounts for the various concentration isolines differ between the various figures. A factor facilitating the comparison is given in the figure legends.

The ground concentration c_0 in the case of the tree margin absorbing all air admixtures shows a pronounced maximum on the windward side of the tree (Fig. 3.23a). This is caused by the retardation of the horizontal wind and the

Table 3.5. Input parameters for dispersion calculation

Number of particles	$N = 20\,000$
x-coordinate of source	$x_q = 2$ m
y-coordinate of source	$y_q = 2$ m
z-coordinate of source	$h_q = 4$ m
Time increment	$\Delta t = 0.2$ s

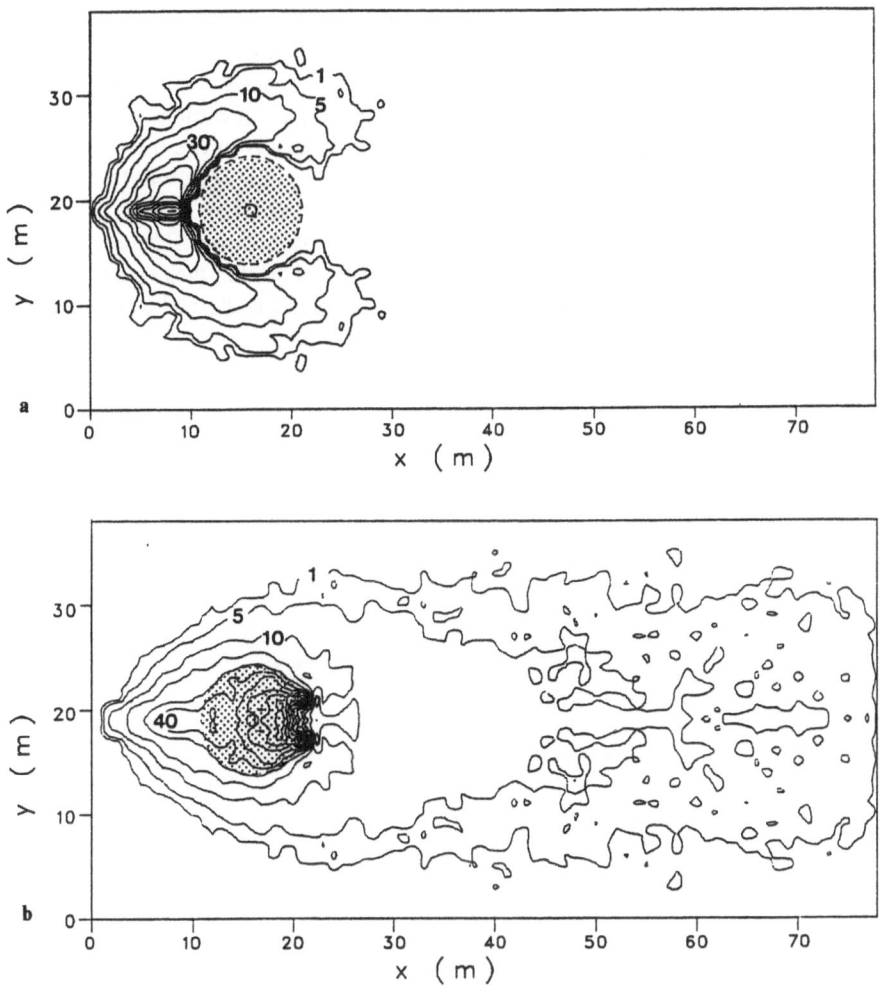

Fig. 3.23. Surface concentration of pollutants for a source height of 4 m and neutral stratification. **a** Cone-shaped tree with total absorption at stand margins; **b** cone-shaped tree with total permeability of the tree. (In both cases the contour values have to be multiplied by 0.1)

ensuing settling. The interior of the tree is almost completely devoid of particles and the concentration of pollutants inside is effectively zero. In this case, the tree acts as an ideal filter. At the margin of the individual tree there is also $c = 0$, a result which at first appears to be surprising as it is here that all particles are filtered out. However, these particles 'disappear' immediately and are no longer present in the volume for which the calculation is carried out. In combination with the flow around the obstacle, this effect leads to the horseshoe-shaped ground concentration, which was also observed by RS-86.

When the particles are allowed to cross the stand without even one of them becoming deposited, a completely different concentration pattern is obtained (Fig. 3.23b). Outside the tree a maximum close to the ground is also found on the windward side of the obstacle. However, it is not as pronounced as in the preceeding simulation. Strong maxima are now observed in the interior and at the rear of the obstacle. As outlined in the preceeding chapters, wind speed and turbulence in the interior of the tree are relatively low. This results in a long residence time of the particles and thus in a high concentration. Though there is advective transport, albeit small, through the obstacle, this dies completely at the rear margin of the stand. The sign of $\bar{u}$ is reversed as here the return eddy

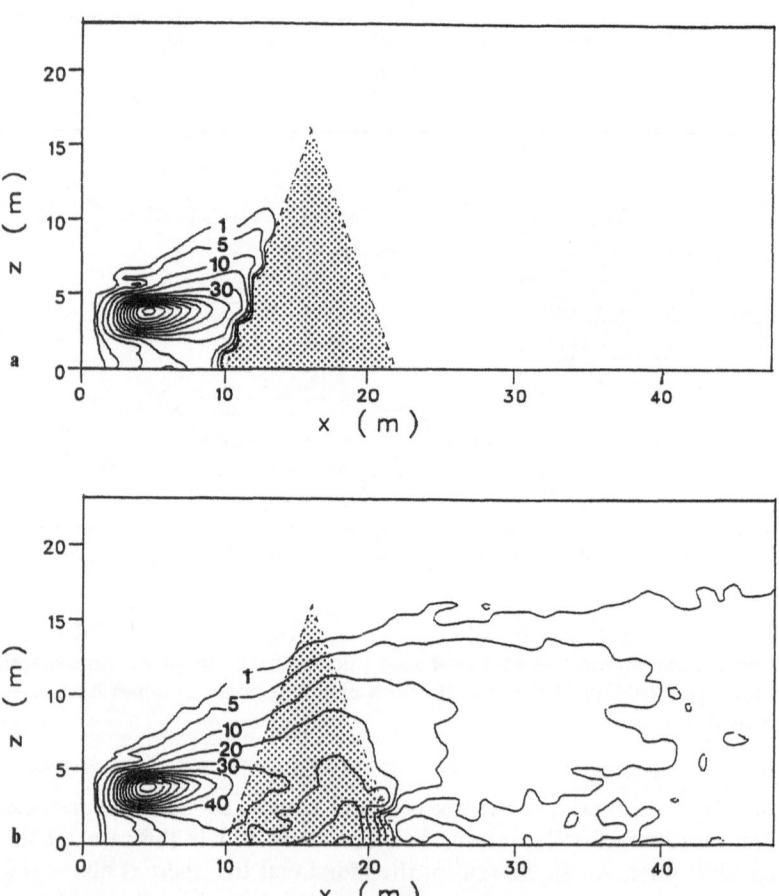

Fig. 3.24. Vertical section of concentration of pollutants at $y = 19$ m and neutral stratification. **a** Cone-shaped tree with total absorption at stand margins; **b** cone-shaped tree with total permeability of the tree. (In both cases the contour values have to be multiplied by 0.1)

starts. Particles reaching this zone need a very long time to escape from it. Mainly the additional turbulent velocity u' is responsible for the reentry of these particles into the outer flow. Along the symmetry plane, the vertical velocity attains values of several cm s^{-1}, sufficient to establish an upward advection. More to the sides, along the lateral margins of the eddy, this mechanism disappears, leading to the two maxima of concentration on the leeward side.

Vertical sections at $y = 19$ m for both simulations illustrate the influence of the different deposition processes. At complete sedimentation many particles are rapidly deposited and are therefore no longer available for further dispersion. As a consequence, the rear end of the obstacle is almost completely devoid of pollutants (Fig. 3.24a).

In the other extreme (Fig. 3.24b), the particles accumulate in zones of low velocity and turbulence, leading to a maximum close to the ground. Otherwise, the pollutant cloud travels through the tree and particles are also encountered far into the lee. The leeside eddy modifies the concentration in such a way that a minimum is simulated close to the ground. This can be ascribed to the fact that the particles have to be carried (horizontally and vertically) in a wide trajectory around the eddy before they come under its direct influence. As this takes some time, particles are increasingly lost and only a few of them reach the zone of reverse flow.

By unrolling the mantle of the cone, it may be shown that total deposition occurs on the margin of the stand. The frontal maximum (0°), i.e., on the side facing into the larger-scale wind, is mainly the result of advection (Fig. 3.25a). This is further indicated by its vertical position, corresponding approximately to the height of the source. On the leeward side (180°) only low concentrations are found, caused by advection and turbulent diffusion.

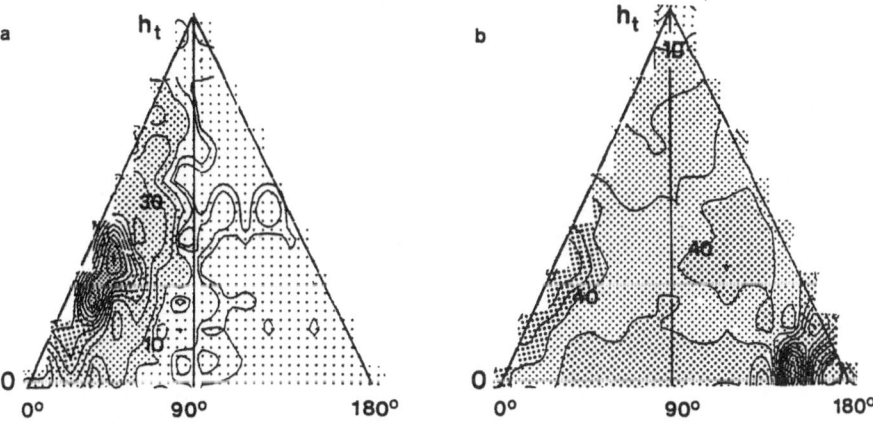

Fig. 3.25. Deposition pattern on unrolled surface of a cone-shaped tree under neutral stratification. **a** Total absorption at stand margins (the contour values as % of the maximum value have to be multiplied by 0.01). **b** Total permeability of the tree (the contour values as % of the maximum value have to be multiplied by 0.1)

In the case of a permeable tree, the maximum in front of the obstacle is also well developed, but the structure at its rear, between 160° and 180° and below 3 m, is more dominant (Fig. 3.25b). This zone is identical to the front margin of the reverse flow area. The reason for this maximum has been explained above.

A lowering of the source height to 2 m results in higher concentrations along the ground. In this case, the pollutant cloud also spreads rapidly upward, as a consequence of the increased turbulence in front of the tree and also of the dynamically induced upward motion (Fig. 3.26a). From the fact that at a source height of 10 m the cloud will not spread upward so much, but will rather spread downward (Fig. 3.26b), it may be concluded that turbulence is, in effect, a dominant factor for the vertical dispersion on the windward side. In both cases, a concentration minimum is found in the reverse flow area behind the tree and

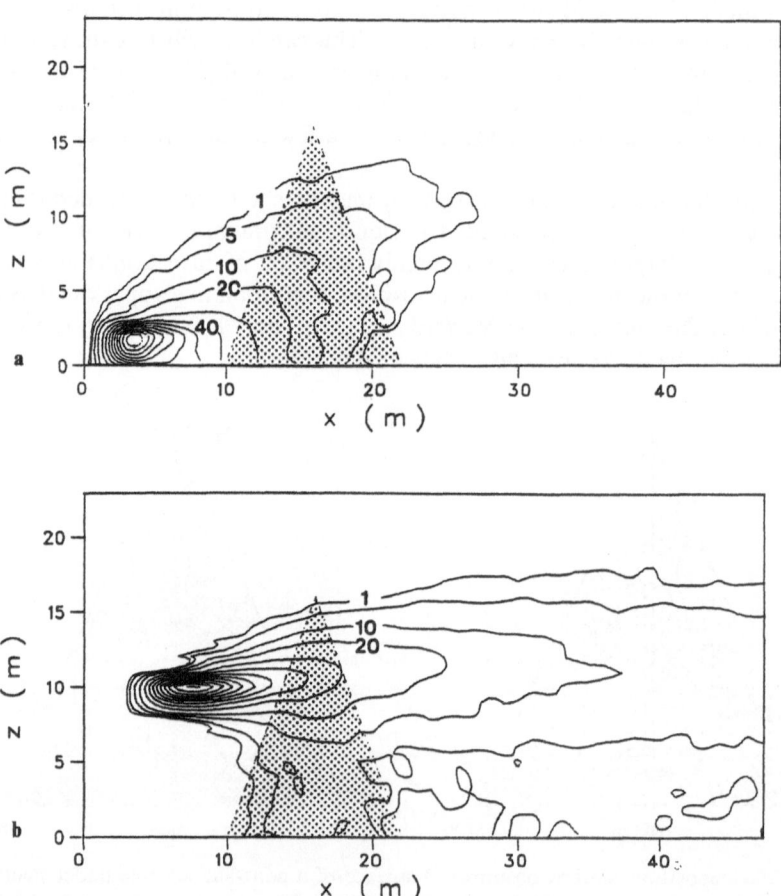

Fig. 3.26. Vertical section of concentration of pollutants for a cone-shaped tree at $y = 19$ m and neutral stratification. **a** Source height of 2 m; **b** source height of 10 m

this is smaller than at $h_q = 2$ m. At a height of 2 m, the particle source is in a region with a very large $\bar{v}$ component, leading to pronounced flow around the obstacle. As a result, the pollutants are led around the tree and enter the leeward eddy rather late and much diluted. In this flow situation only a few particles are transported back to the obstacle and, accordingly, a low concentration is calculated. The situation is different for $h_q = 10$ m at which the $\bar{v}$ component is rather low. Advection of the particles is rather rapid due to the relatively high transport velocity in the x-axis and particles have almost no chance of becoming mixed down to the ground because of the elevated turbulence on the leeward side of the tree. As a result, the concentration along the ground is generally very low.

Unrolling the mantle surface in these two cases, it is again found that the positions of the concentration maxima at the margin of the tree on the windward side correspond to the source heights (Fig. 3.27). In both cases there is also a leeward, albeit somewhat lower maximum near the ground which coincides with the margins of the reverse flow zone.

In a stable stratified atmosphere, turbulence will be lower and the pollutant plume will not spread as far as under neutral conditions. This is especially evident on the leeward side of the tree (Fig. 3.28).

Similar air pollution studies were carried out for other tree shapes. The results will be explained for the simulation of an ellipsoidal-shaped tree with a 3-m trunk under neutral stratification. Because of the underflow now possible and the resulting strong downward motion in front of the tree, the plume axis is deflected downward (Fig. 3.29). The relatively high wind speed in the trunk zone and the control of the air flow by the shape of the obstacle inhibit the dispersal of

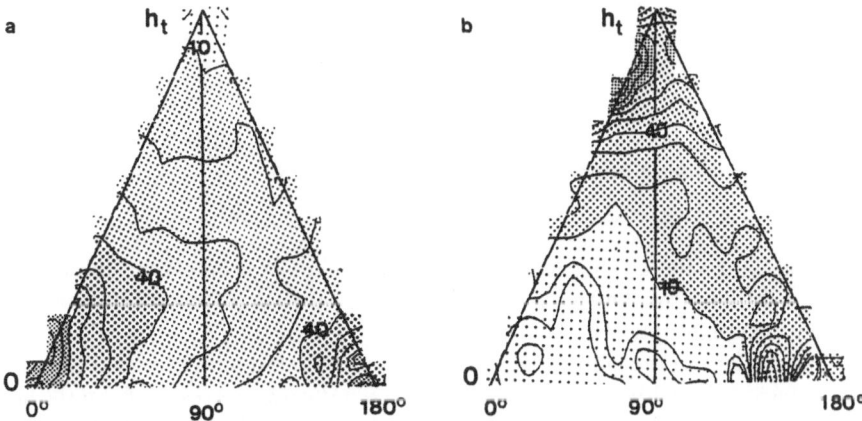

Fig. 3.27. Deposition pattern on unrolled surface of a cone-shaped tree under neutral stratification. **a** Source height of 2 m (the contour values as % of the maximum value have to be multiplied by 0.01). **b** Source height of 10 m (the contour values as % of the maximum value have to be multiplied by 0.1)

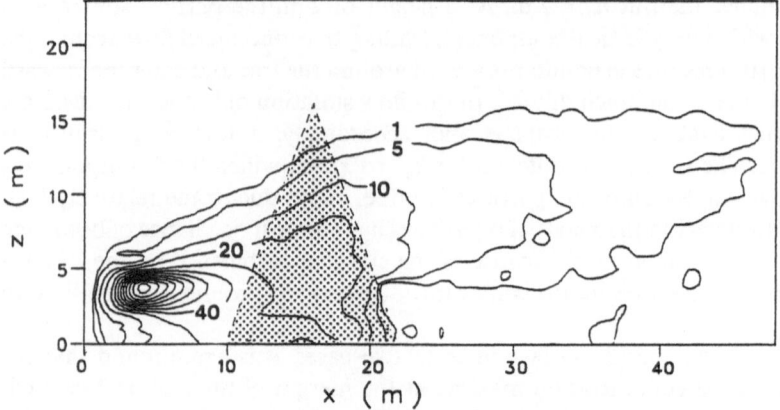

Fig. 3.28. Vertical section of concentration for a cone-shaped tree at $y = 19$ m and a source height of 4 m (stable stratification).

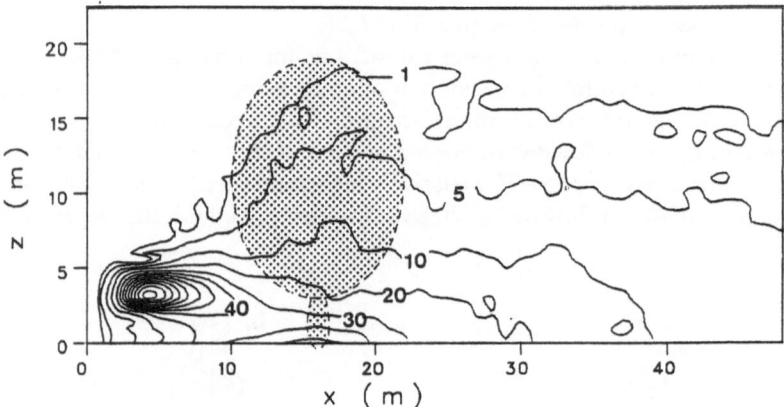

Fig. 3.29. Vertical section of concentration of pollutants for an ellipsoidal-shaped tree at $y = 19$ m and a source height of 4 m (neutral stratification)

the pollutant cloud. This leads to relatively high concentrations along the ground. In front of the tree and in its interior, turbulent processes cause the plume to spread upward.

The surface of the ellipsoidal tree shows one maximum at the front and one at the back (Fig. 3.30). The maximum on the windward side results from the fact that this position is located just opposite the source. As outlined above, the pollutant cloud will not spread as much when underflow occurs and the concentration on the rear side attains a relatively high value. Because of the upward motion in this zone, the particles are carried upward to a considerable extent and lead to a secondary maximum on the leeward side of the tree.

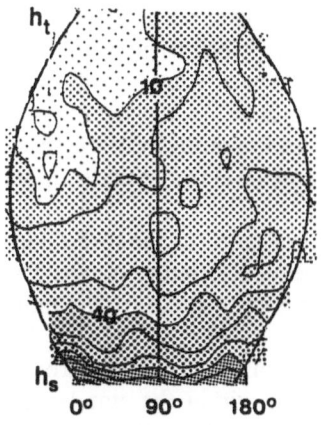

Fig. 3.30. Deposition pattern on unrolled surface of an ellipsoidal-shaped tree under neutral stratification (the contour values as % of the maximum value have to be multiplied by 0.1)

In agreement with theoretical approaches (Friedlander and Johnstone 1957; Rouhiainen and Stachiewicz 1979; Trela 1982) and wind tunnel experiments (RS-86), the numerical simulations also show that at low velocities deposition of pollutants will take place mainly in zones of high turbulence intensity.

3.3 Studies on Groups of Trees

In a further study the wind and turbulence patterns around groups of trees were investigated. The groups consist of ellipsoidal trees varying in number and location.

In all simulations with isolated trees, acceleration takes place along the sides of the obstacles and retardation occurs on the leeward side; these effects are also observed in nature (Woelfle 1937). Quite different phenomena, however, may be observed when the wind penetrates a forest from the surrounding open ground. Because of the width of the forest lateral flow, as occurs around an isolated tree, is not possible and the complete advancing air mass has either to enter the forest or to flow over it. The portion of the wind penetrating the forest is rapidly subjected to strong retardations, caused by the dissipation of kinetic energy due to friction on the trunks in the trunk zone and on branches and leaves in the canopy. As a consequence, the air quickly calms down as it advances farther into the stand. This convergence of the air mass in the vicinity of the forest edge leads to a stagnation which extends into the open ground. Thus, the wind has already been slowed down to some degree ahead of the obstacle. In a horizontally homogeneous stand, this convergence leads to an upward flow both in and above the canopy as the necessary mass balance has to be maintained.

Meroney (1968) carried out wind tunnel experiments to investigate the changes in wind and turbulence profiles during transition from the open ground into a stand. The experimental forest consisted of 19-cm-high plastic trees with

6-cm-high trunks, randomly distributed over an area 2×11 m^2. The logarithmic wind profile present outside the stand became increasingly deformed as the edge of the forest was approached. Within the canopy lower wind speeds were observed, declining almost completely in the crown zone. In the trunk zone, a maximum was developed as a result of the jet effect. Quantitatively the speed of this jet was below the velocity observed at the same level in the undisturbed profile.

Visualizing flow by smoke, a pronounced rise was noted at the forest edge and this may be ascribed to the horizontal convergence referred to above. The distribution of that part of the smoke penetrating the forest is controlled more by diffusion than by advection.

The low values of the longitudinal turbulence intensity over open ground further decreased upward. In the transition zone and within the stand itself, a pronounced turbulence maximum was observed slightly above the top of the crowns. Within the trunk zone it attained rather low values which were smaller than at the same level over open ground.

Nägeli (1954, cited in Mitscherlich 1971) investigated changes in wind speed in front of, within and behind an approximately 550-m-wide pine stand. A retardation of the wind was already present in front of the forest at a distance of ten times the tree height. In the marginal zone, the jet effect leads to an acceleration of the flow which is, however, dissipated fairly rapidly. Thereafter, the wind slows down almost exponentially with increasing advance into the interior of the forest. In the lee of the forest, the velocity notably increases again beyond a narrow calm zone. But, even at a distance of 20 times the tree height, the flow has not yet reestablished the original undisturbed pattern that was prevalent prior to entry into the forest.

The wind patterns observed by Nägeli may be considered as generally applicable. The values themselves, however, are strongly controlled by the stand structure of the forest. Geiger (1926) and Geiger and Amann (1931) described large differences in wind velocities between a pine stand with or without undergrowth and an oak forest.

The forest edge is especially exposed to wind damage (Mitscherlich 1973; Mayer 1985). The flow patterns prevalent in this zone have been described already. As the flow is slowed considerably in the lower portion of the atmospheric boundary layer by the stand elements, a corresponding acceleration has to take place above the trees. Lateral flow effects will not be taken into account here. This leads to higher wind loading of individual trees that are more exposed and which may break as a result. In this context it has been reported by Woelfle (1937) that during storms the first few rows of trees frequently remain intact, while more or less damage occurs only farther into the forest.

In the first row of trees to be investigated, three identical ellipsoidal trees with a diameter of 10 m, crown height of 16 m and trunk height of 3 m were arranged behind one another in the direction of the wind. The trees were spaced in such a way that there were no gaps between the crowns.

In Fig. 3.31a the $\bar{u}$ component of the velocity is shown for a vertical section along the symmetry plane. The penetration of the wind into the trunk zone, associated with a velocity maximum at the front, is clearly recognizable. Within the crowns the velocity is low and reverse flow occurs in the lee of each crown. There is now a zone of weak backflow close to the ground behind the stand, quite in contrast to the situation for the isolated tree. Above the obstacle, velocity is increased, as is to be expected from the required mass conservation. It is significant that this maximum does not develop directly at the edge of the stand, but rather over the second and third trees. This is in agreement with the field observations on wind damage in forests. One explanation is that for the first tree there is still a rather intense underflow, whereas air around the trunk

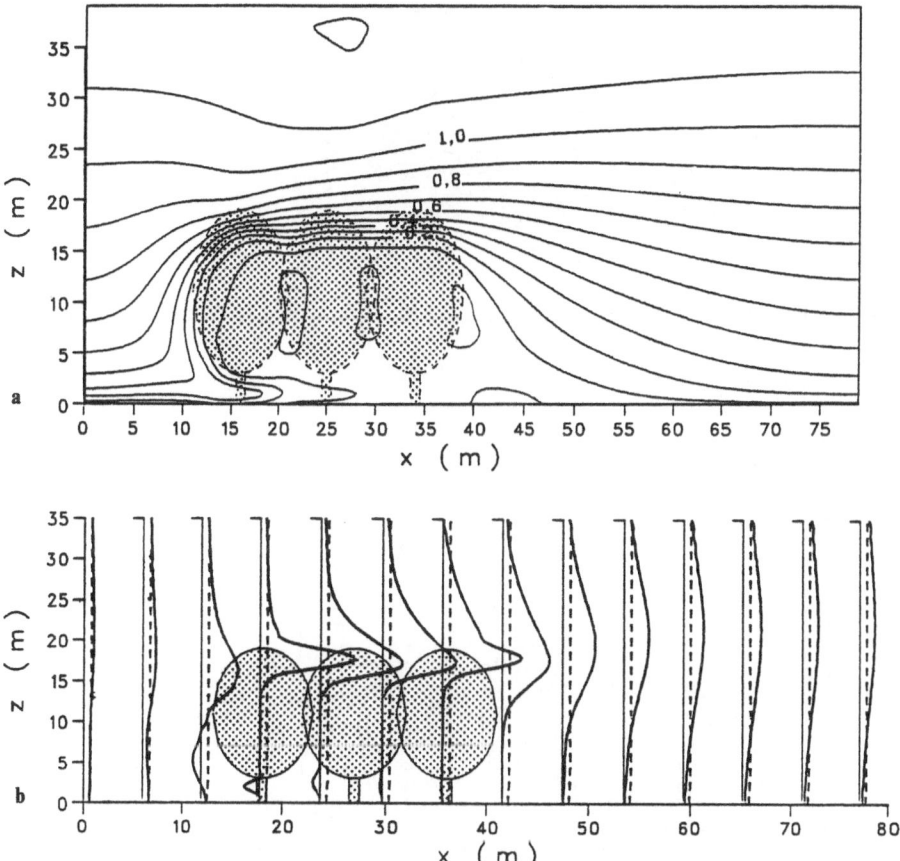

Fig. 3.31. a Vertical section of $\bar{u}$ component at $y = 35$ m for air flow over row of trees arranged parallel to the direction of $\bar{u}(H)$ (m s^{-1}, intervals 0.1 m s^{-1}). **b** Vertical profiles of shear stress at $y = 35$ m (undisturbed profile is shown by *broken line*)

zone of the other two trees is almost completely calm. When the integrated vertical mass flux at each horizontal location along this section is assumed to be constant, then the wind over the more distant trees has to be correspondingly greater. Lateral flow effects can be neglected as the $\bar{v}$ component is zero along the symmetry plane.

The turbulence structure, represented by the vertical turbulent momentum flux (Fig. 3.31b), shows the same results as the wind tunnel experiments. The pronounced maximum in the upper part of the canopy contrasts with the lower

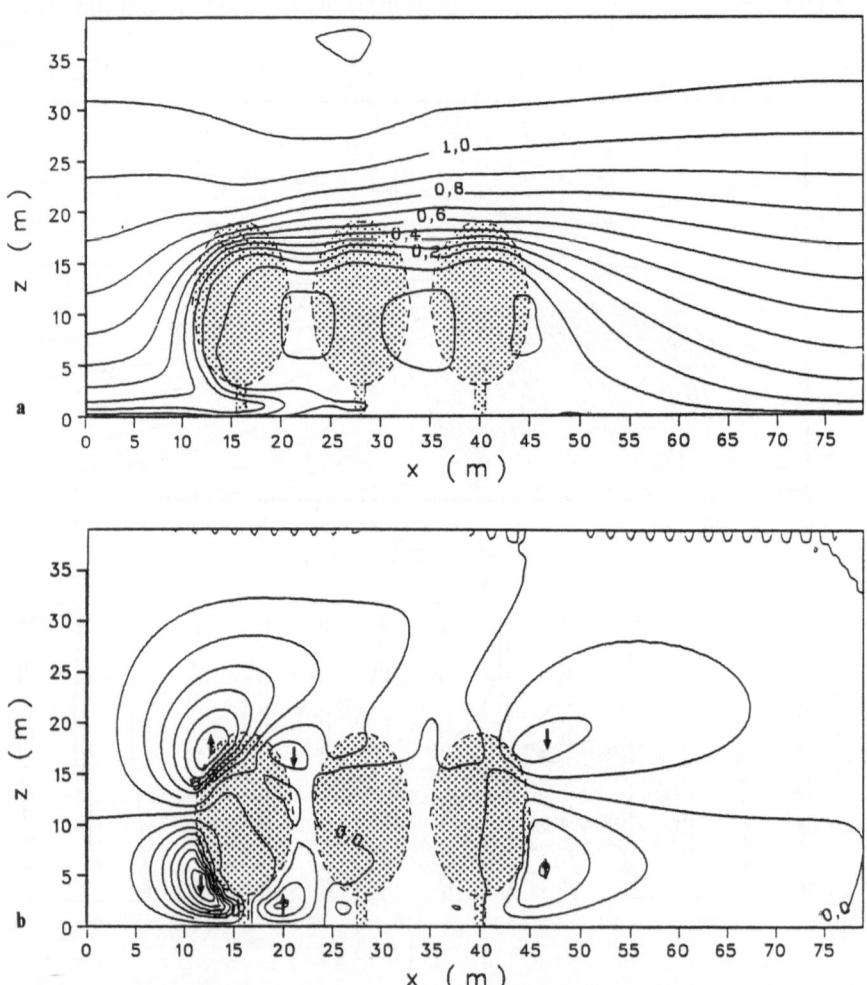

Fig. 3.32. Vertical sections for air flow over row of trees arranged parallel to the direction of $\bar{u}(H)$. **a** $\bar{u}$ component (m s^{-1}, intervals 0.1 m s^{-1}); **b** $\bar{w}$ component (cm s^{-1}, intervals 2.5 cm s^{-1}). *Arrows* indicate upward and downward motion

values within the rest of the stand. Farther beyond, the elevated maximum becomes gradually dissipated by mixing with the overlying and underlying air masses. It is, however, still recognizable even at the end of the model domain.

When the distance between the crowns is increased and a gap free of stand elements is established, the reverse flow zone between the individual trees becomes wider (Fig. 3.32a). At the same time, penetration of the outer flow into the canopy increases. The velocity maximum in the trunk zone, the leeward eddy and the maximum above the second tree all remained the same as in the preceeding case.

The distribution of the vertical velocity in this vertical section (Fig. 3.32b) shows the pronounced rise on the windward side as was found in the wind tunnel experiment. The strong downward motion is also observed and this causes advection of momentum into the trunk zone, leading to the velocity maximum there.

Beyond the first tree, the horizontal convergence leads to an upward motion close to the ground and to a weak downward motion in the upper part of the crown. Only in the lee of the total stand are distinct patterns observed and these are similar to those which occur during flow around an isolated tree. The quantitative differences of the vertical motions, arranged in a cloverleaf pattern behind and in front of the obstacle, are rather small when the tree row is compared with an isolated ball-shaped tree with a trunk. In another stand configuration, the three ellipsoidal-shaped trees were arranged not in line, but next to one another at right angles to the wind, leading to a relative wider obstacle than considered so far.

While flow around the isolated tree showed a maximum of $\bar{v} = 13 \text{ cm s}^{-1}$, the value was trebled in this simulation. However, even this relatively strong $\bar{v}$ component was not sufficient to lead the complete air mass entering the inflow boundary around the obstacle. This is only possible when there is an increased $\bar{u}$ component in the parts of the field not occupied by stand elements. Following some slowdown in front of the obstacle, the wind is strongly accelerated in the trunk zone, leading to values which are double those encountered at the same height over open ground. A pronounced maximum is developed, especially where the crowns of adjacent trees are in the contact with each other, i.e., where the widest openings are present close to the ground. In the lee of the trunks, the wind shadow effect is clearly recognizable (Fig. 3.33a).

The flow, which is led in a wide circle around the obstacle, forms a zone of reverse flow far in the lee of the tree (Fig. 3.33b) and this appears to owe its presence to the conservation of mass. The $\bar{v}$ component leads to a strong convergence and thus to a higher pressure in this zone which, in turn, results in retardation and acceleration.

The region with $\bar{u} < 0$ acts like an extension of the stand and it is only behind this zone of reverse flow that the leeward downward motion, which is usually observed directly behind the stand, sets in (Fig. 3.33c). The relatively wide obstacle also influences the vertical velocity, the extreme values of which are effectively doubled.

76   Air Flow Around and Through Individual Trees

In a further simulation, the distance between the individual trees was enlarged so that flow between the trees became possible. This immediately reduces the tendency of the flow to evade the obstacle. The most pronounced difference from the previous calculations is that there is no leeward zone of reverse flow (Fig. 3.34). However, in roughly the same position a minimum is found for the $\bar{u}$ component which still attains about 25% of the value over open ground. In addition to rows of trees arranged parallel to or at right angles to the main wind direction, the investigation also covered groups of trees. The group consisted of three rows, each of four identical trees, and this represented a considerable obstacle to the air flow. Flow above and under the trees was correspondingly larger in the case of the crowns being in close contact with each other. The values were $\bar{v}_{max} = 38$ cm s^{-1} and $\bar{w}_{max} = 36$ cm s^{-1}.

The length of the leeward zone of reverse flow also depends on the width of the obstacle. It is thus not surprising to note that the area with $\bar{u} < 0$ extends beyond the outflow boundary (Fig. 3.35a). The eddy pair in the lee of the group

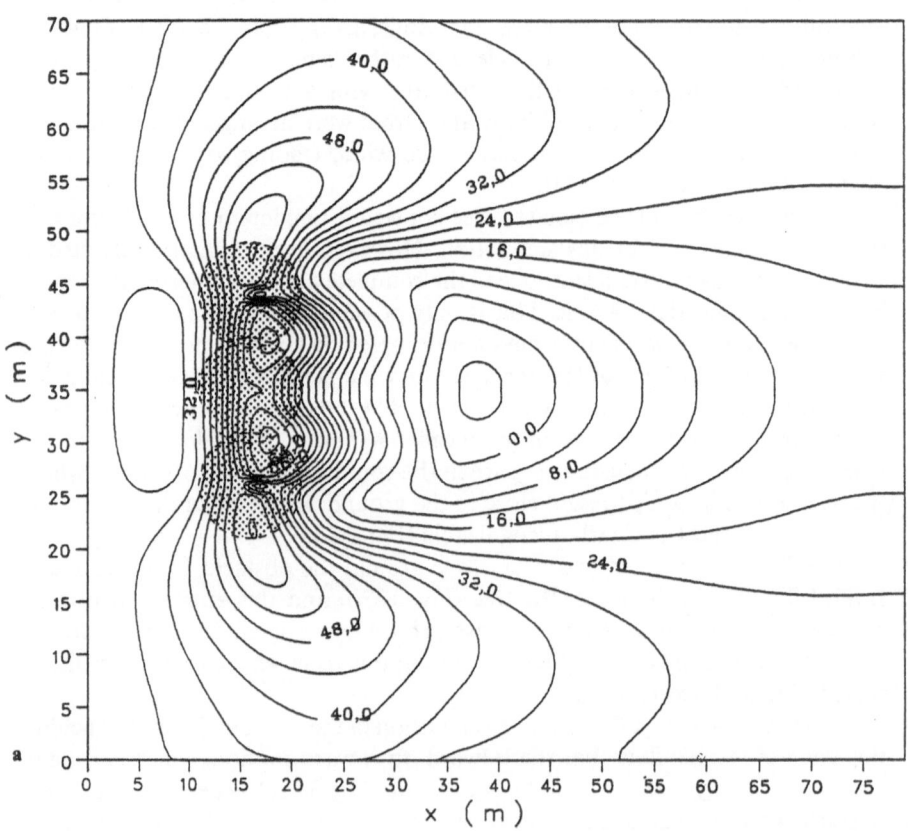

Fig. 3.33a

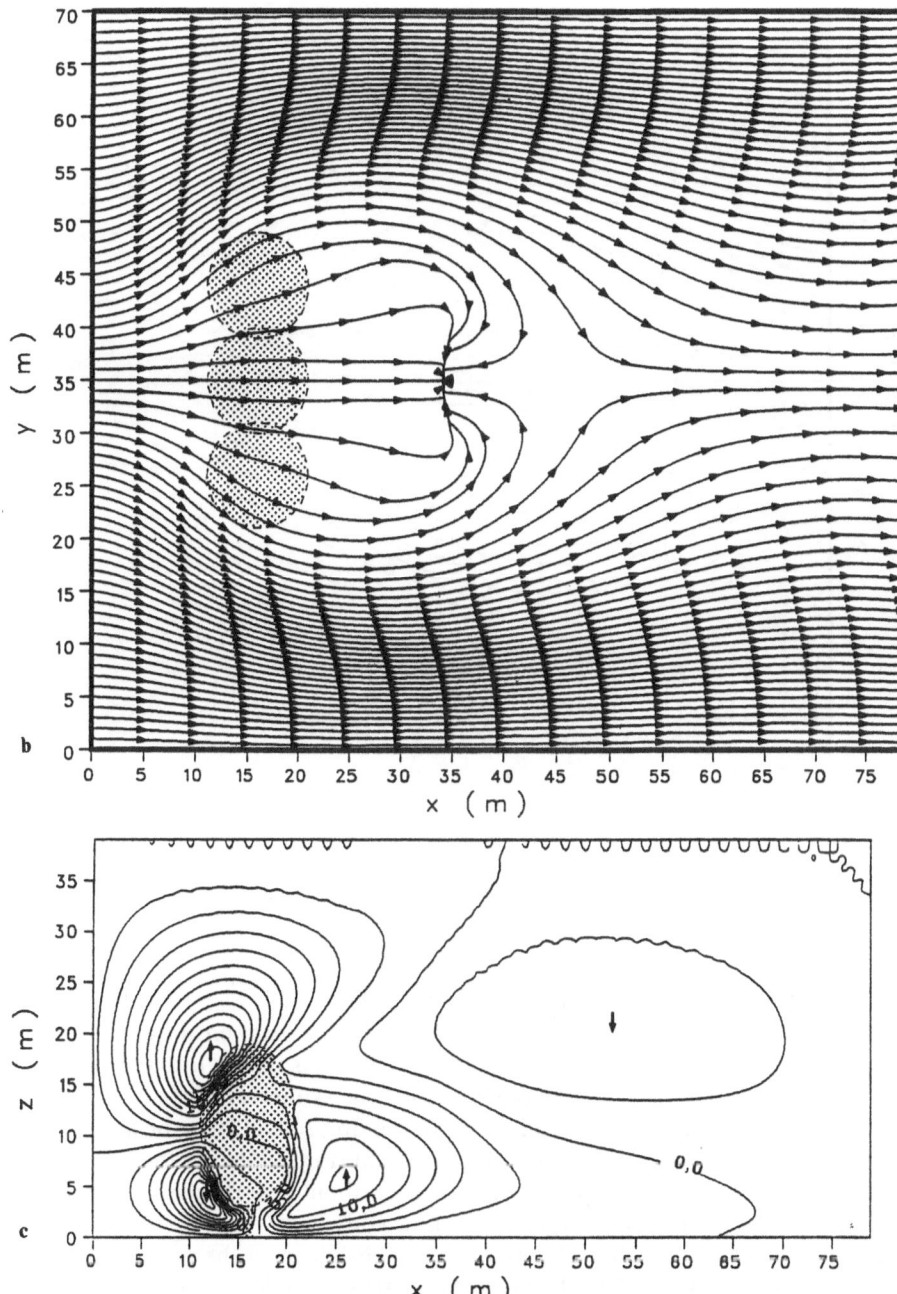

Fig. 3.33. Air flow over row of trees arranged at right angles to $\bar{u}(H)$ for neutral stratification. **a** Horizontal section of $\bar{u}$ component (cm s^{-1}, intervals 4 cm s^{-1}); **b** streamlines 1 m above ground; **c** vertical section of $\bar{w}$ component (cm s^{-1}, intervals 2.5 cm s^{-1})

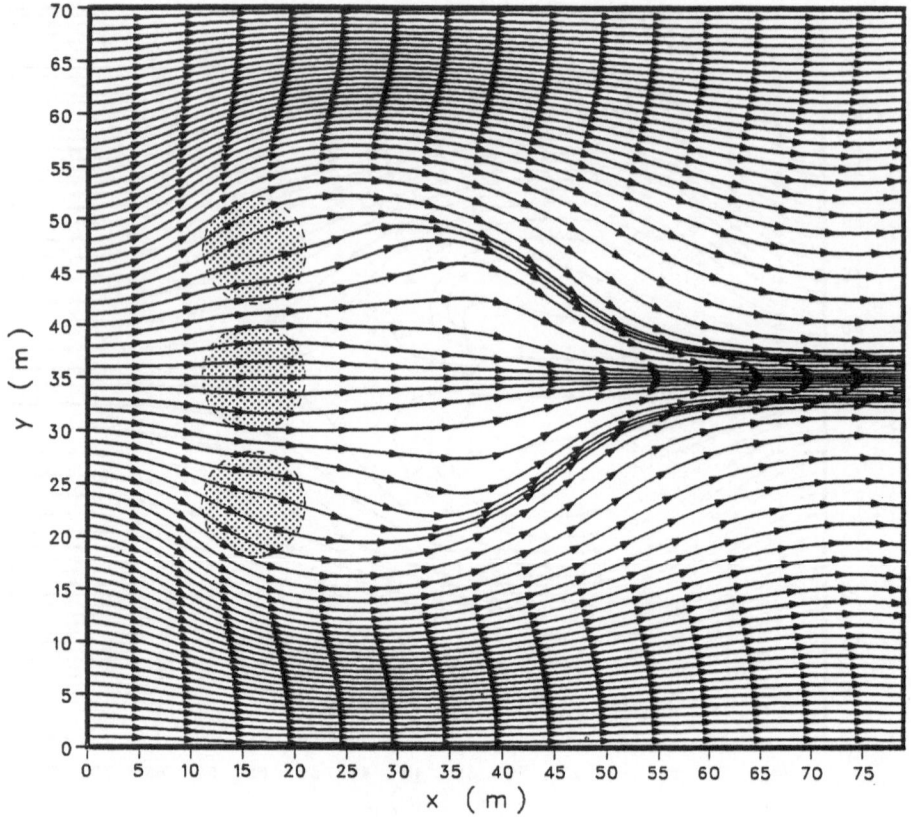

Fig. 3.34. Streamlines 1 m above ground for air flow over row of trees arranged at right angles to $\bar{u}(H)$ for neutral stratification

is the most spectacular feature in this flow pattern. It is also present behind isolated trees, but not to the same degree as here. The diagram shows that flow still takes place through the stand, even though it has only a low velocity.

Vertical profiles of the $\bar{u}$ component along the symmetry plane again show the velocity maximum in the trunk zone of the first row of trees; however, this becomes rapidly dissipated farther into the stand (Fig. 3.35b). At the end of the stand the air is almost completely calm up to about two-thirds of the crown height. Directly adjacent, a large reverse flow eddy is developed which, even at the outflow boundary, still inhibits the reestablishment of the original profile. The structures of the profiles shown here for various positions in the stand agree well with the results of the wind tunnel experiments by Meroney (1968).

As outlined above, the velocity maximum close to the ground within the forest may be approximated rather well by an exponential function (Nägeli 1954; Flemming 1964). This situation is also observed in the present simulations

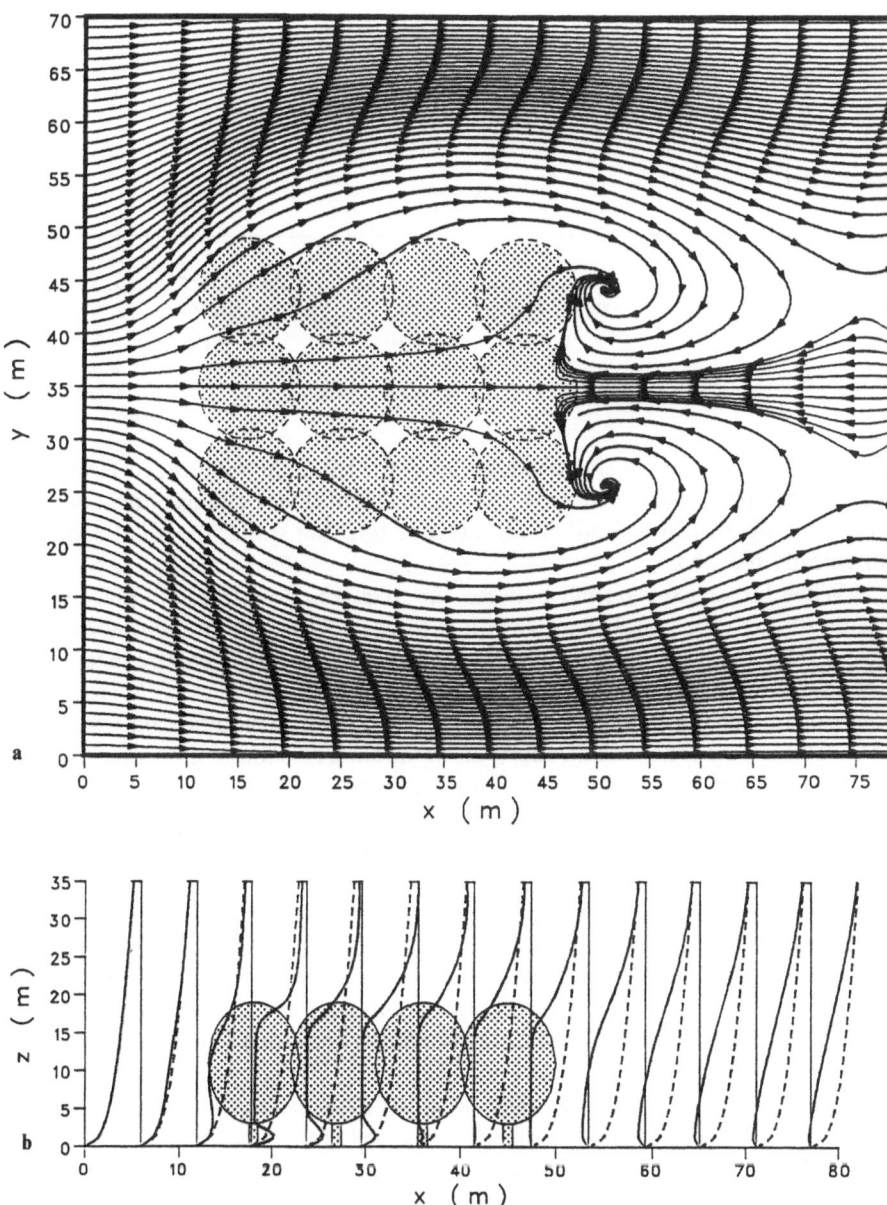

Fig. 3.35. Air flow over a group of 12 ellipsoid-shaped trees for neutral stratification. **a** Streamlines 1 m above ground; **b** vertical profiles of u component at $y = 35$ m (undisturbed profile is shown by *broken lines*); **c** normalized horizontal profile of $\bar{u}$ component 1 m above ground at $y = 35$ m. *Shaded area* indicates width of the obstacle

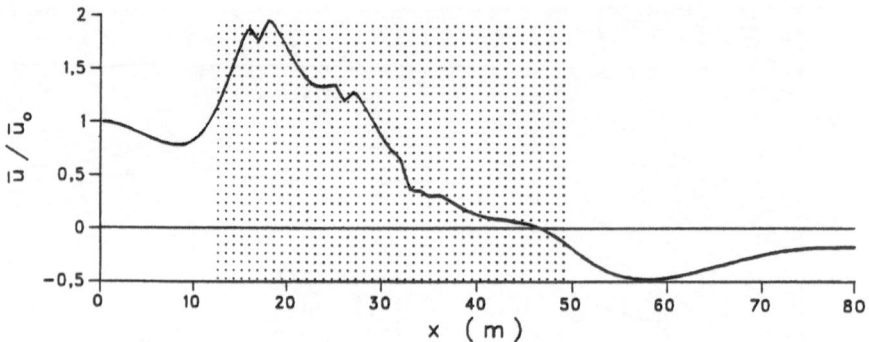

Fig. 3.35c

(Fig. 3.35c). Following a slowdown ahead of the trees, the wind is accelerated in the trunk zone of the first tree row to such an extent that it attains almost double the value over open ground. With advance into the stand, the wind speed is drastically reduced.

When larger gaps are left between the trees, the flow can take place more easily through the stand than when it is forced to flow around and above the stand. This becomes immediately obvious from the velocity components since the maximum is reduced to $\bar{v}_{max} = 29$ cm s^{-1} and $\bar{w}_{max} = 27$ cm s^{-1}. However, the general structure of the horizontal velocity field is modified only slightly. The reverse flow zone with an eddy pair in the lee is also found in this simulation (Fig. 3.36a). Because it is wider and longer, as a result of the known relation of its areal extent with the geometric dimensions of the total obstacle, the total length of the zone can only be guessed as the zone with $\bar{u} < 0$ here also extends beyond the outflow of the field.

The flow contours passing the first row of trees show an interesting feature. Flow takes place not only around the obstacle as a whole, but also around the individual trees. Each tree shows a leeward zone in which the $\bar{u}$ component disappears almost completely, while the $\bar{v}$ component, and with it the vertical velocity, attain considerable values. This effect is underlined by the sharp bends in the streamlines.

It was observed by RS-86b that along the gaps in the stand parallel to the direction the superimposed wind $\bar{u}(H)$, the velocity is much higher than in a vertical section through the trees themselves. Vertical sections along the symmetry plane (section A, Fig. 3.36b) and along the gap between the first and second row of trees (section B, Fig. 3.36c) show that in the latter case the outside flow penetrates far into the stand. Again, the maximum of the horizontal velocity is found at a height of 35 m above the second tree. This has been outlined already in the discussion of the susceptibility of stand margins to wind damage.

The velocity at 1 m above ground level is generally larger in section B (Fig. 3.36d). In section A the acceleration in the trunk zone of the first two

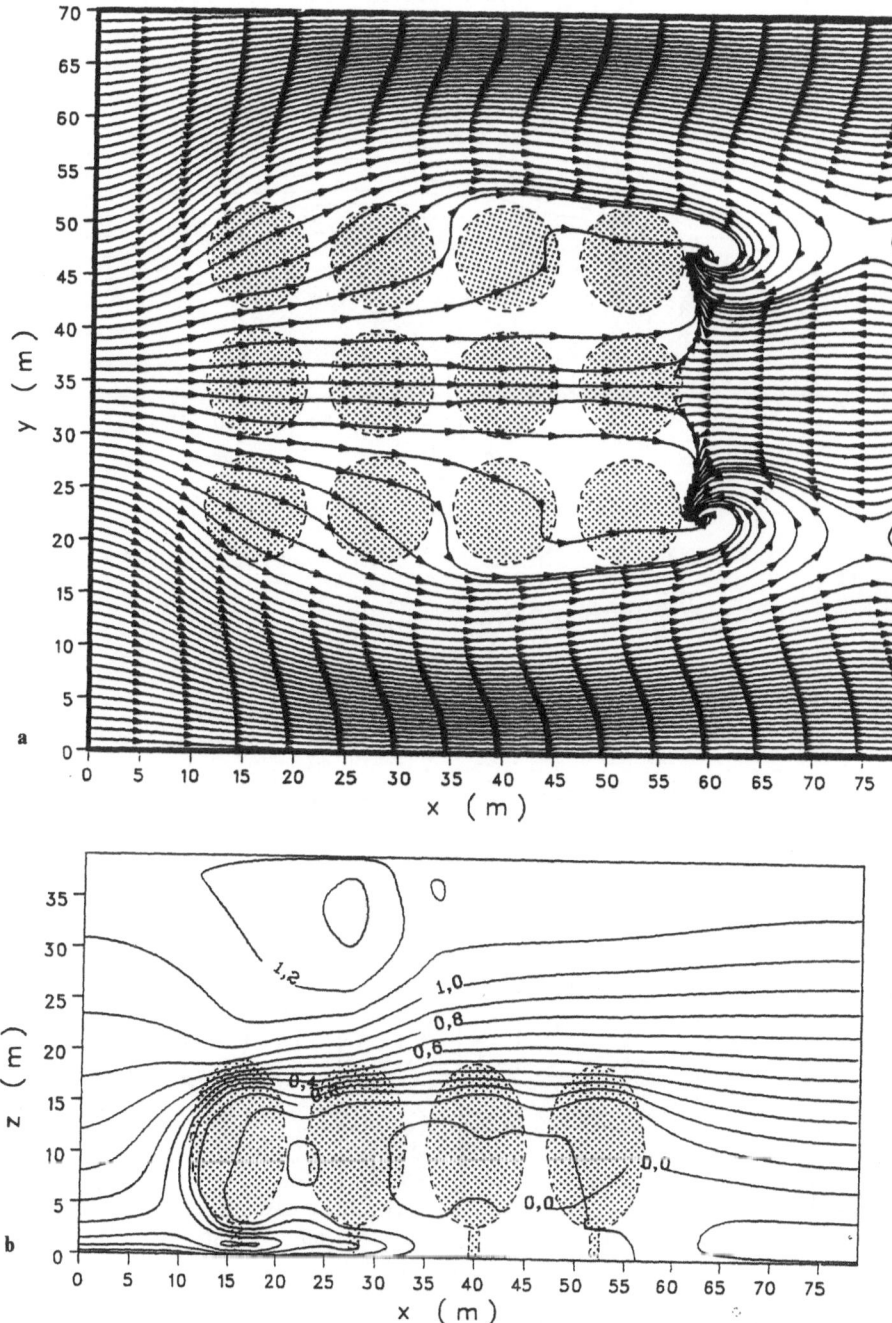

Fig. 3.36. Air flow over a group of 12 spaced ellipsoid-shaped trees for neutral stratification. **a** Streamlines 1 m above ground; **b** vertical section (referred to as section A) of $\bar{u}$ component at $y = 35$ m

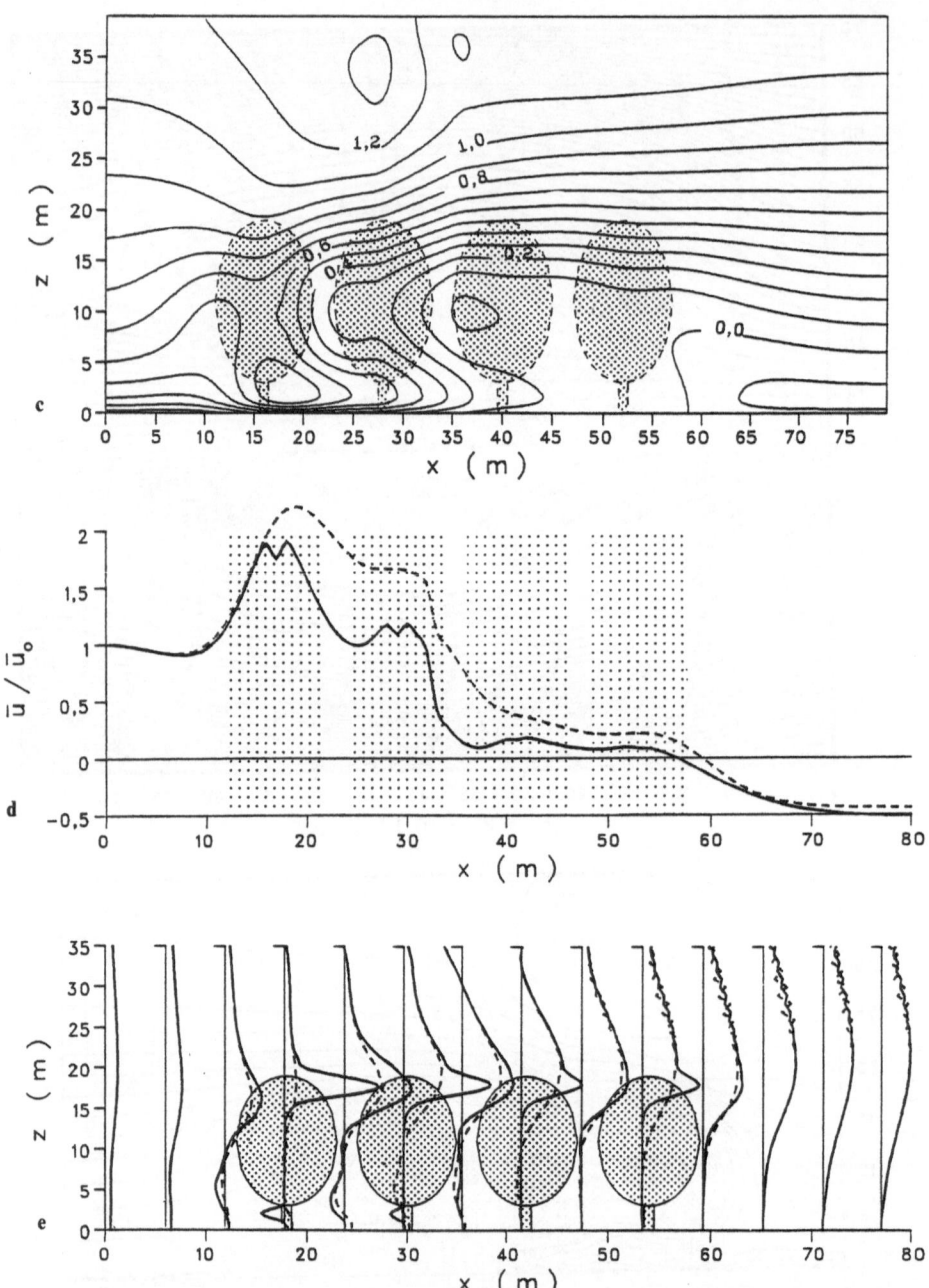

Fig. 3.36 (*cont.*). **c** vertical section (referred to as section B) of $\bar{u}$ component at $y = 29$ m; **d** horizontal profiles of $\bar{u}$ component 1 m above ground (*solid line* section A; *dashed line* section B; *shaded areas* indicate the width of the obstacle); **e** vertical profiles of shear stress (*solid line* section A; *dashed line* section B)

trees becomes clearly evident. The maximum of the turbulence parameters observed by Meroney (1968) and RS-86 in the upper part of the canopy is also simulated in the present calculations (Fig. 3.36e). In section B, between the trees this parameter shows a similar trend despite the absence of obstacles.

3.4 Investigations of Wind Shelter Hedges

In many regions windbreaks and shelterbelts are used to reduce soil erosion and to improve plant growth. Usually a long narrow obstacle is oriented perpendicular to the main wind direction. Shape and permeability (porosity) vary within wide limits from rows of dense tall trees and straw mats to densely braided fences.

The variety of obstacles used clearly illustrates that the question as to which type of windbreak results in an optimum effect has not yet been resolved. This is particularly difficult as a large number of input variables have to be considered and these may add to or subtract from each other in effect. A detailed summary of such investigations was presented by van Eimern et al. (1964).

The wind shelter effect of an obstacle is usually given as the percentage of velocity reduction normalized to the value over open ground.

It is generally held that although the air flow is strongly retarded in the lee of a dense obstacle, the undisturbed wind speed is reestablished rather rapidly. For small porosities the degree of reduction is less pronounced, but is still felt at some distance beyond the obstacle (Nägeli 1946; van Eimern et al. 1964). Wilson (1985) could not confirm this statement during investigations on modern wind shelter fences. The largest velocity reduction, together with a wider sheltered zone, was found behind denser fences.

A problem which is repeatedly encountered is that of the determination of the porosity. Whereas the porosity of artificial fences is determined fairly easily, so far a suitable method has not been found for natural growth like trees or shrubs. The optimum porosity values found in the literature (e.g., $P = 0.5$ in Baltaxe 1967) are of only indicative value for the design of wind shelter hedges.

The sheltering effect is also controlled by wind speed and wind direction. Van Eimern (1957) noted that the velocity reduction in the lee of a 12-m-tall maple tree stand bordering a road was less for higher velocities than for lower velocities. Results by Nägeli (1946) support these observations while the opposite situation was reported by Kittridge (1948). At an oblique impact angle of up to 45° against the maple-tree row the reduction was virtually unchanged; however, Nägeli (1965) observed a pronounced reduction in similar experiments.

Thermal stratification of the atmosphere close to the ground leads to a lower sheltering effect during stable conditions than during unstable daytime conditions (van Eimern 1957).

When several obstacles are arranged behind one another a larger area may be protected (Woodruff 1956; Nägeli 1965). In this case the intervals between the rows of obstacles have to be selected in such a way that the open ground wind speed is not achieved again between the rows.

The shape and arrangement of windbreak strips are controlled largely by local conditions, and recourse to results from other sites is possible only to a small extent.

In the following it will be shown that numerical simulation models can also be applied to these problems.

As the sheltering hedges are very long obstacles, it is possible to use a two-dimensional version of the model. The obstacle studied is a 1.5-m-wide and 2-m-high row of bushes with a permeability of 50%. In order to resolve such a small obstacle, the grid spacing in the horizontal and vertical directions was taken as 0.5 m. The upper boundary was set at 30 m, i.e., 15 times the obstacle height. In the reference run, an undisturbed wind speed of $\bar{u}(H) = 2 \text{ m s}^{-1}$ was assumed in a neutral stratified atmosphere.

In Fig. 3.37 the calculated horizontal and vertical wind components are shown in vertical cross sections. Around the obstacle the air flow has been greatly slowed down and already on the windward side a sheltering effect, albeit

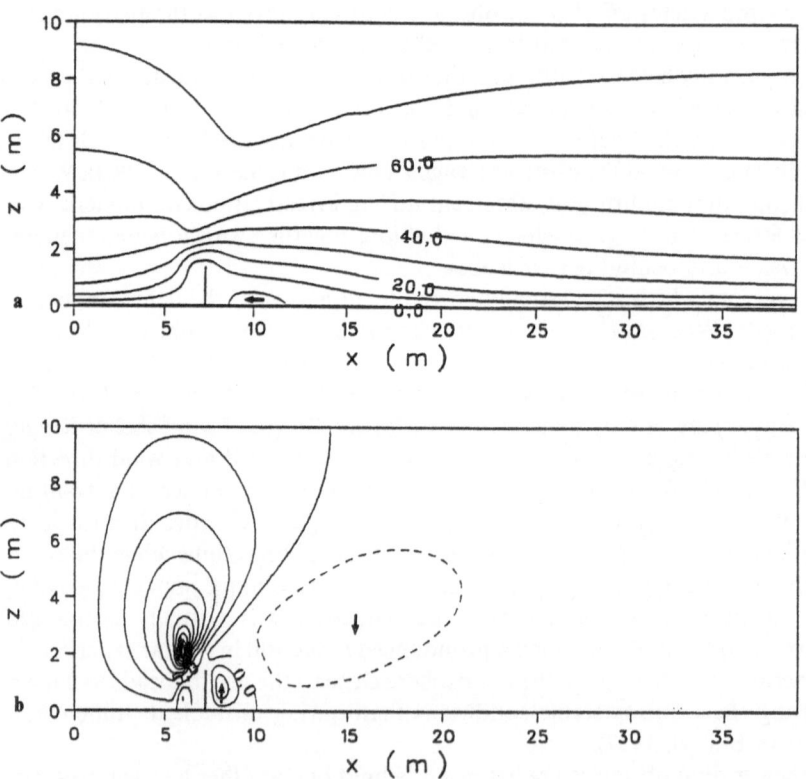

Fig. 3.37. a, b

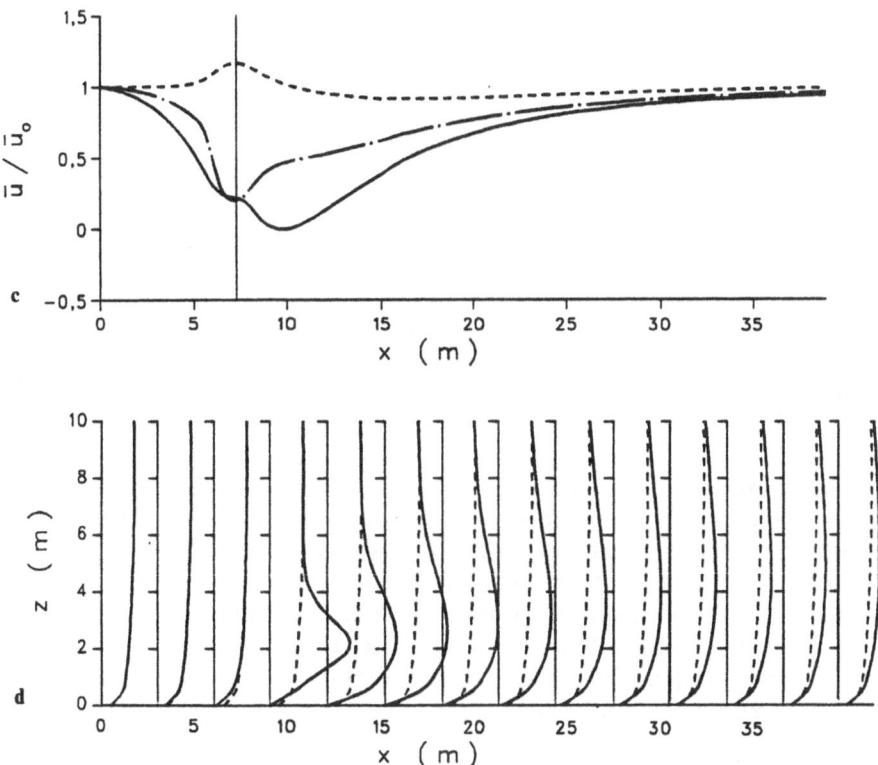

Fig. 3.37. Air flow over a shelterbelt (2 m high, 1.5 m wide, $P = 0.5$). **a** Vertical section of $\bar{u}$ component [in % of $\bar{u}(H)$]; **b** vertical section of $\bar{w}$ component (cm s^{-1}, intervals 2.5 cm s^{-1}, *arrows* indicate upward and downward motion); **c** normalized horizontal velocity profiles (*solid line* $z = 0.25h_t$, *dashed* and *dotted line* $z = 0.75h_t$, *dashed line* $z = 2h_t$); **d** vertical profiles of normalized shear stress (undisturbed profile is shown by *broken lines*)

of limited extent, may be noted. At some distance beyond the hedge a weak leeward eddy is obtained with a velocity of less than 1% of the undisturbed value at the same elevation over open ground. The strong reduction of $\bar{u}$ close to the ground leads to the transport of a correspondingly larger air mass over the obstacle. This is possible only with an increased velocity in this region as indicated by the simulated values above the hedge. Due to the relatively high vertical velocity, the advected air is deviated upward by the obstacle, to return to the original position only far on the leeward side. Similar flow patterns including the leeward minimum, were noted by Nägeli (1953) during investigations of the wind field in the vicinity of a 2.2-m-high reed curtain.

Horizontal velocity profiles at different heights, normalized to the undisturbed value, illustrate the extent of the sheltering effect (Fig. 3.37c). At a height

$z = 0.25h_t$ a minimum is found for $\bar{u}$ at a distance of $x = 1.5h_t$, even at a distance of $x = 10h_t$ only 80% of the undisturbed velocity has been reattained. At $z = 0.75h_t$ the length of the sheltered zone is considerably reduced and the lowest velocity is found within the hedge iteself.

The normalized vertical shear stress profiles (Fig. 3.37d) in this simulation also show good agreement with patterns observed in wind tunnel experiments and field measurements. The strong wind shear in the lee of the hedge leads to the elevated maximum. Above a height of $z = 5h_t$, the calculations show only minor deviations from the undisturbed profile, while Bradley and Mulhearn (1983) observed a height of $z = 4h_t$.

When calculations are carried out with hedges of different porosity it is not possible to confirm the observations of, e.g., Nägeli (1946). The leeward velocity minimum and the length of the sheltered zone become more pronounced with the density of the hedge (Fig. 3.38). However, it was possible to confirm the observations of Wilson (1985).

A fence or a wind shelter hedge may also exert negative influences. Over sloping terrain and during the night cold air drainage sets in and this may be blocked by such an obstacle (King 1973). This results in lower temperatures on the upslope sides and therefore to an increase in the possibility of night frost. An increase in the permeability of the fence reduces this effect but, at the same time, diminishes its actual sheltering purpose. Under these circumstances it may be better to vary the porosity of the fence with height. Then, the greater permeability will be arranged closer to the ground to permit the desired drainage of cold air.

For a hedge with a mean porosity of $P = 0.7$ three simulations were carried out. In the first, a uniform porosity was assumed, whereas in the second the lower half possessed a higher density than the upper half. In the third simulation the situation was reversed. The evaluation shows that the leeward sheltering

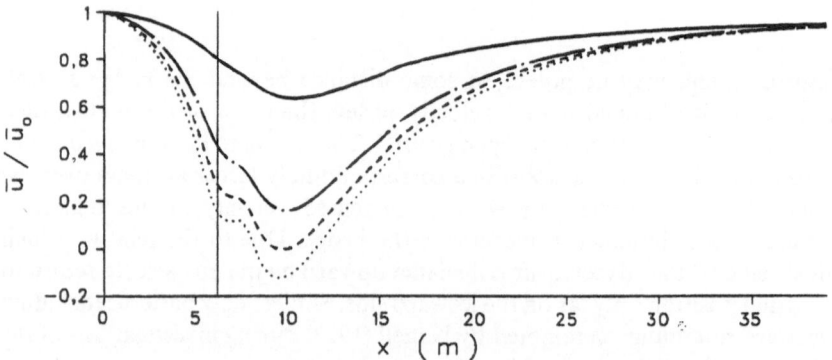

Fig. 3.38. Normalized horizontal velocity profiles at $z = 0.25h_t$ for different porosities. *Solid line P = 0.9; dashed and dotted line P = 0.7; dashed line P = 0.5; dotted line P = 0.3*

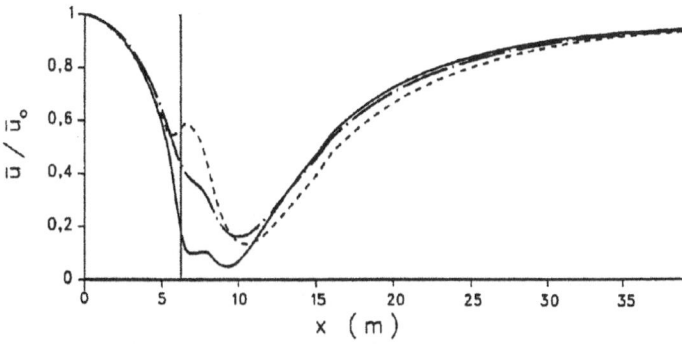

Fig. 3.39. Normalized horizontal velocity profiles at $z = 0.25h_t$ for different porosity patterns. *Solid line* Bottom denser; *dashed line* top denser; *dashed* and *dotted line* homogeneous density

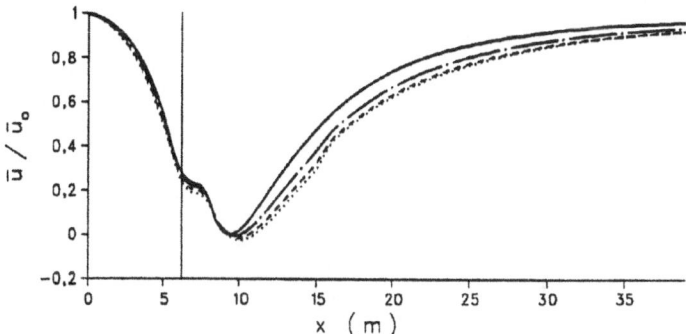

Fig. 3.40. Normalized horizontal velocity profiles at $z = 0.25h_t$ for different superimposed wind speeds $\bar{u}(H)$. *Solid line* $1\,\mathrm{m\,s}^{-1}$; *dashed* and *dotted line* $2\,\mathrm{m\,s}^{-1}$; *dashed line* $4\,\mathrm{m\,s}^{-1}$; *dotted line* $8\,\mathrm{m\,s}^{-1}$

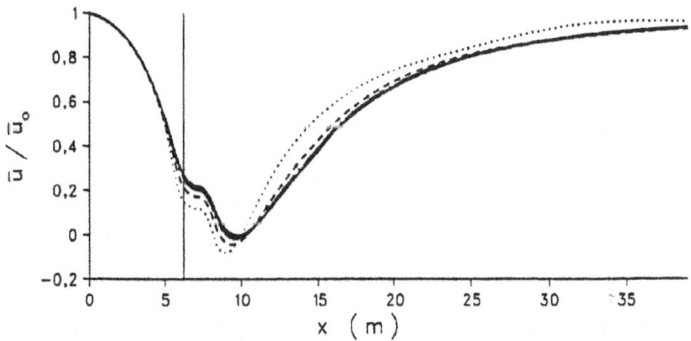

Fig. 3.41. Normalized horizontal velocity profiles at $z = 0.25h_t$ for different wind directions. *Solid line* $90°$; *dashed* and *dotted line* $67.5°$; *dashed line* $45°$; *dotted line* $22.5°$

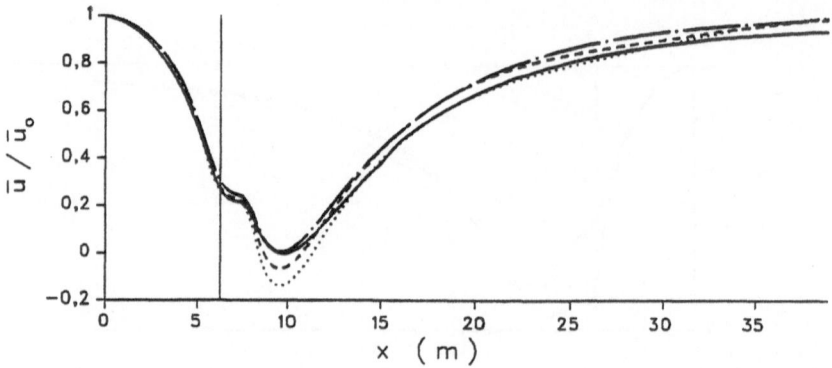

Fig. 3.42. Normalized horizontal velocity profiles at $z = 0.25h_t$ for different thermal stratifications. *Solid line* 0 K/100 m; *dashed* and *dotted line* 0.35 K/100 m; *dashed line* 1 K/100 m; *dotted line* 2 K/100 m

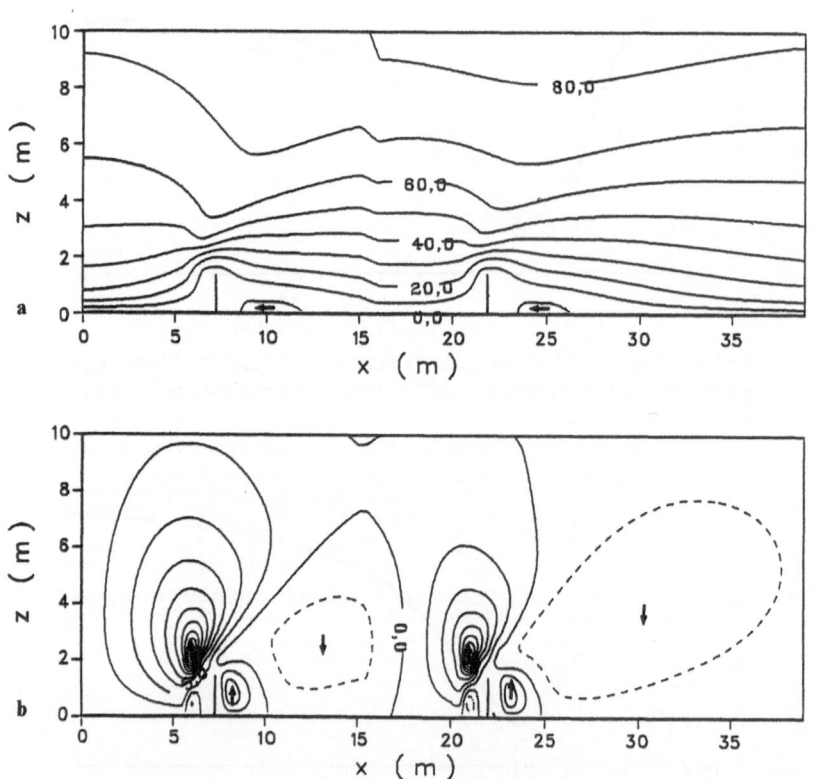

Fig. 3.43. Vertical sections for air flow over two consecutive shelterbelts. **a** $\bar{u}$ component [in % of $\bar{u}(H)$; *arrows* indicate reverse flow]; **b** $\bar{w}$ component (cm s^{-1}, intervals 2.5 cm s^{-1}, *arrows* indicate upward and downward motion)

effect was larger when the obstacle was less dense close to the ground (Fig. 3.39). A similar result was found by Wilson (1987) for a 1.12-m-high porous fence.

The relationship between wind strength and sheltering effect of a fence has not yet been clearly resolved. As shown by the review of van Eimern et al. (1964), a number of authors observed a more pronounced sheltering effect with increasing wind speed, while others found exactly the opposite. Simulations with four different velocities ranging from 1 to 8 $\mathrm{m\,s^{-1}}$ result in a decrease in the leeward minimum of $\bar{u}$ and in an extension of the sheltered zone for larger wind speeds (Fig. 3.40). However, the differences are fairly small. At $\bar{u}(H) = 8\ \mathrm{m\,s^{-1}}$, 80% of the undisturbed velocity is encountered again at a greater distance than at $\bar{u}(H) = 1\ \mathrm{m\,s^{-1}}$, the interval corresponding to about twice the obstacle height.

The sheltering effect is reduced as the impact angle of the wind becomes smaller (Fig. 3.41). At identical wind speeds a horizontal rotation of the wind leads to a decrease in the wind component perpendicular to the hedge and, as outlined above, to a narrower sheltered zone.

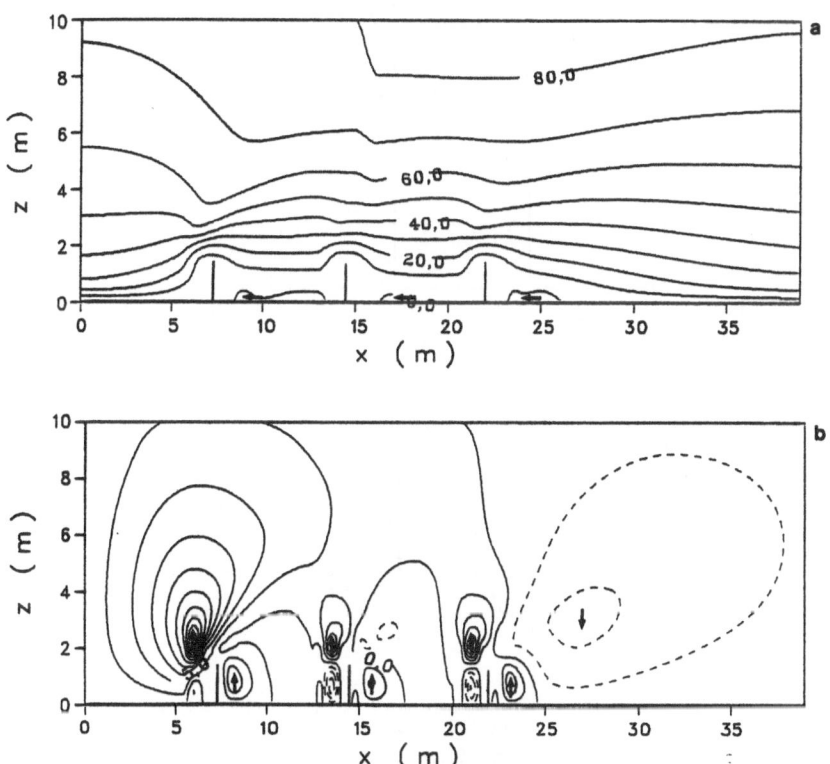

Fig. 3.44. Vertical sections for air flow over three consecutive shelterbelts. **a** $\bar{u}$ component [in % of $\bar{u}(H)$, *arrows* indicate reverse flow]; **b** $\bar{w}$ component (cm s^{-1}, intervals 2.5 cm s^{-1}, *arrows* indicate upward and downward motion)

Furthermore, it was possible to confirm the observations of van Eimern (1957) that with increasing stability the reduction of $\bar{u}$ in the lee extends over a smaller area (Fig. 3.42). The minimum is more pronounced under very stable stratification, but rapidly reapproaches the original value. When evaluating this presentation it should be borne in mind that the near-surface wind speeds of the various simulations vary considerably. The undisturbed velocity over open ground is much lower in the stable case than under neutral stratification, and thus a reverse flow zone is more easily established behind the fence.

An individual wind shelter hedge on a plain acts like an obstacle against the oncoming wind. Several such rows arranged behind one another may be interpreted as providing an increased surface roughness at which the air flow tries to establish a new equilibrium. In the interval between two hedges separated by a distance of $8h_t$, the air flow is again able to penetrate down to the ground and the $\bar{u}$ and $\bar{w}$ fields in the immediate vicinity of each obstacle show a structure that is similar to the one developed along a single hedge (Fig. 3.43). If

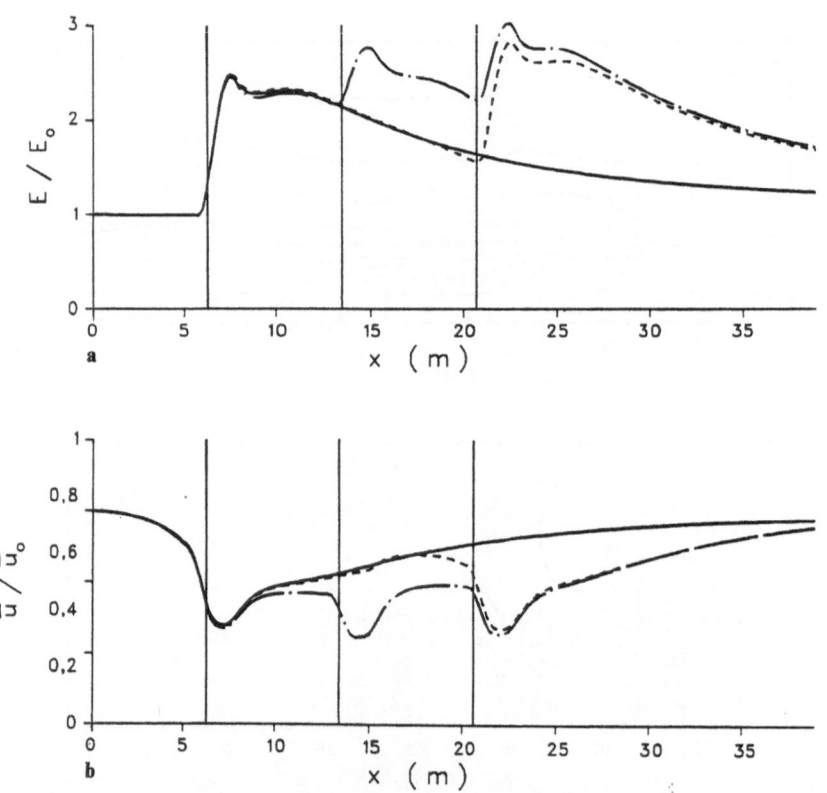

Fig. 3.45. Normalized horizontal profiles at $z = 0.75h_t$ for spaced obstacles. **a** Turbulent kinetic energy; **b** velocity. *Solid line* 1 hedge; *dashed line* 2 hedges; *dashed* and *dotted line* 3 hedges

another hedge is placed between the two initial ones, an almost completely calm zone is simulated (Fig. 3.44). The strong upward motion in front of the first hedge leads the air flow in a wide trajectory over the obstacle as a whole.

Investigations by Nägeli (1965) on similar groups of obstacles showed that in the lee of the wider spaced hedges almost the same degree of reduction of $\bar{u}$ was observed as behind the closer spaced ones. As Fig. 3.45a shows, the simulations yielded rather similar results. It is interesting to note that the width of the sheltered zone is the same regardless of whether two or three hedges are considered in the simulations, but the zone is longer than in the case of an isolated hedge. The turbulent kinetic energy at $z = 0.75h_t$ shows a pronounced maximum at each obstacle, the intensity of the maxima increasing with each additional obstacle (Fig. 3.45b). This latter situation may be explained by advection, resulting from larger values of E produced on the windward side of the hedges.

4 Air Flow Through and Above Stands

In Chapter 3 the wind patterns around individual trees, either isolated or arranged in groups, were investigated. The shape of the trees was represented explicitly with a total of about 2500 grid points for each obstacle.

However, in simulations where in addition to the effects of the stand on the distribution of meteorological variables the effects of topography will also be investigated, the horizontal grid intervals must be enlarged to 100–1000 m. This implies that taller stands may be considered in the model only through a parameterization and not in detail. For a realistic simulation the influence of slope inclination, season, etc. on the wind pattern has to be taken into account. The model used so far is not suitable for this purpose and must be extended accordingly.

4.1 Improvement of the Numerical Model

The investigation of the three-dimensional distribution of the meteorological variables under the influence of stand and topography is facilitated by the mesoscale simulation model FITNAH (flow over irregular terrain with natural and anthropogenic heat sources). The calculations are carried out on a numerical grid with the centers representing values for the box volume $\Delta x \, \Delta y \, \Delta z$. For each area $\Delta x \, \Delta y$, a degree of forest density, n_c, is defined and this indicates the percentage of area covered by forest and open ground.

As the individual variables differ notably between the forest (index c) and the open ground (index f), a representative value for a certain grid point is found by weighting. When such a variable is designated as ϕ, there is

$$\phi = (1 - n_c) \cdot \phi_f + n_c \cdot \phi_c. \tag{4.1}$$

This relation will be used later in the derivation of the final equation system.

In complex terrain the use of the common Cartesian coordinate system is sometimes not recommended, as the position of a grid center will not necessarily coincide with the topographical height. As all model variables are only defined on grid points, it will be difficult to find suitable boundary conditions, especially at ground level. Although this problem can be avoided with the aid of special techniques (Mason and Sykes 1979; Ulrich 1987), the irregularly shaped integration area is usually transformed into a rectangular area through a respective transformation relation. For this purpose a new vertical coordinate z^* is defined in such a way that the ground coincides with a plane $z^* = $ constant (Gal-Chen and Sommerville 1975; Becker 1978). However, this transformation has the disadvantage that the coordinate system is no longer orthogonal and that additional metric terms appear in the model equations.

The transformation relation is

$$z^* = \frac{z - h(x, y)}{H - h(x, y)},\tag{4.2}$$

with h being the orography and H the height of the upper boundary of the model. In Fig. 4.1 the position of various planes $z^* = $ constant is shown in a vertical section in the Cartesian coordinate system.

The transformation of the model equations to the (x, y, z^*) coordinate system used here was described by, e.g., Dutton (1976); Pielke (1984) and Gross (1984).

After transformation of the model equations into the terrain following coordinate system and considering the canopy effects [Eq. (3.8)], the system of equations reads as follows:

$$\frac{du}{dt} = f(v - v_g) - f^* w - \frac{1}{\varrho}\frac{\partial p'}{\partial x} + \frac{z^* - 1}{H - h}\frac{\partial h}{\partial x}\frac{1}{\varrho}\frac{\partial p'}{\partial z^*} + \frac{\partial}{\partial x}\left(2K_{mx}\frac{\partial u}{\partial x}\right)$$

$$+ \frac{\partial}{\partial y}\left(K_{my}\frac{\partial u}{\partial y}\right) + \frac{1}{(H - h)^2}\frac{\partial}{\partial z^*}\left(K_{mz}\frac{\partial u}{\partial z^*}\right) - n_c c_d b u V,\tag{4.3}$$

$$\frac{dv}{dt} = -f(u - u_g) - \frac{1}{\varrho}\frac{\partial p'}{\partial y} + \frac{z^* - 1}{H - h}\frac{\partial h}{\partial y}\frac{1}{\varrho}\frac{\partial p'}{\partial z^*} + \frac{\partial}{\partial x}\left(K_{mx}\frac{\partial v}{\partial x}\right)$$

$$+ \frac{\partial}{\partial y}\left(2K_{my}\frac{\partial v}{\partial y}\right) + \frac{1}{(H - h)^2}\frac{\partial}{\partial z^*}\left(K_{mz}\frac{\partial v}{\partial z^*}\right) - n_c c_d b v V,\tag{4.4}$$

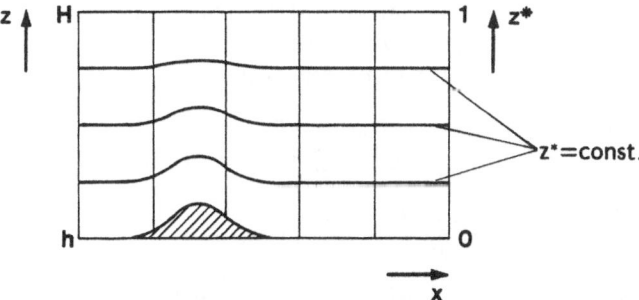

Fig. 4.1. Terrain-following coordinate system. The topography is *shaded*

$$\frac{dw}{dt} = f^*u - \frac{1}{\varrho}\frac{1}{H-h}\frac{\partial p'}{\partial z^*} + g\frac{\theta'}{\tilde{\theta}} + \frac{1}{H-h}\frac{\partial}{\partial x}\left(K_{mz}\frac{\partial u}{\partial z^*}\right)$$

$$+ \frac{1}{H-h}\frac{\partial}{\partial y}\left(K_{mz}\frac{\partial v}{\partial z^*}\right) - n_c c_d bwV, \tag{4.5}$$

$$0 = \frac{\partial}{\partial x}[\varrho(H-h)u] + \frac{\partial}{\partial y}[\varrho(H-h)v] + \frac{\partial}{\partial z^*}[\varrho(H-h)w^*] \tag{4.6}$$

with

$$\frac{d}{dt} = \frac{\partial}{\partial t} + u\frac{\partial}{\partial x} + v\frac{\partial}{\partial y} + w^*\frac{\partial}{\partial z^*}. \tag{4.7}$$

The bar, which indicates the average of the variables over the volume $\Delta x\,\Delta y\,\Delta z$ in the foregoing chapters, will be omitted in the following discussion.

In the z^* system an equation is usually solved for the vertical velocity w^*. As this procedure would entail the solution of a rather elaborate equation due to the multitude of transformation terms (cf. Gross 1984), the third equation of motion for w is solved in the z^* system. The vertical velocity w^*, which is also required, is obtained from:

$$w^* = \frac{w}{H-h} - \frac{1-z^*}{H-h}\left(u\frac{\partial h}{\partial x} + v\frac{\partial h}{\partial y}\right). \tag{4.8}$$

The larger-scale pressure gradient in Eqs. (4.3) and (4.4) has been replaced by the geostrophic wind relation.

The mesoscale pressure disturbance p' is split, according to Patrinos and Kistler (1977), into a hydrostatic part p'_h and a dynamic portion p'_d. The former may be calculated from:

$$\frac{1}{H-h}\frac{\partial p'_h}{\partial z^*} = g\varrho\frac{\theta'}{\tilde{\theta}}. \tag{4.9}$$

After splitting the pressure, as outlined above, Eq. (4.9) is subtracted from the third equation of motion [Eq. (4.5)]. This eliminates a possible source of error as the rather large value of the thermodynamic portion is compensated completely by the pressure p'_h. As a consequence, the influence of the other terms, which are usually one order of magnitude smaller, is observed much more readily in the numerical solution. The equation for the turbulent kinetic energy is:

$$\frac{dE}{dt} = \frac{K_{mz}}{(H-h)^2}\left[\left(\frac{\partial u}{\partial z^*}\right)^2 + \left(\frac{\partial v}{\partial z^*}\right)^2\right] - \frac{K_{hz}}{H-h}\frac{g}{\tilde{\theta}}\frac{\partial\theta}{\partial z^*} - \frac{E^{1.5}}{a^{-3}l} + \frac{\partial}{\partial x}\left(K_{hx}\frac{\partial E}{\partial x}\right)$$

$$+ \frac{\partial}{\partial y}\left(K_{hy}\frac{\partial E}{\partial y}\right) + \frac{1}{(H-h)^2}\frac{\partial}{\partial z^*}\left(K_{hz}\frac{\partial E}{\partial z^*}\right) + n_c c_d bV^3. \tag{4.10}$$

The vertical diffusion coefficient for momentum, K_{mz}, is calculated from E according to Eq. (3.12). l is the Blackadar mixing length, and the effect of stratification on the factor a is provided by Eqs. (3.14)–(3.16).

Provided that the temperature within a given stand volume does not change, the energy budget

$$R_N = Q_H + Q_V + Q_B + Q_D + Q_P + Q_M \tag{4.11}$$

has to be balanced.

The term Q_D is usually neglected, not so much because it may be too small, but rather because it is very difficult to determine. Thom (1975) estimated that Q_D is usually in the range $1–10 \, W \, m^{-2}$, values as high as $100 \, W \, m^{-2}$ being the exception. In the latter case it is inadmissible to neglect the divergence of the horizontal flux of sensible and latent heat. As this will occur only in extreme cases and as there is no general relation for the determination of Q_D, this term will not be considered here.

Q_P is treated in the same way. Although there are frequently large temperature differences between a trunk and the air surrounding it, heat storage in or transfer from the timber stock of a stand is not taken into account. The determination of Q_P is influenced by such a large number of parameters that it is nearly impossible to incorporate this term in the model in a realistic way. These parameters include orientation of the trunk in relation to the sun, nature of the trunk surface, trunk diameter, type of tree, water supply from the soil, etc. Observations by, e.g., Stewart and Thom (1973) show that Q_P may be neglected for low vegetation, while in tall forest stands values of $20–60 \, W \, m^{-2}$ can be achieved. At certain times of day this flux may exceed the latent and sensible turbulent heat fluxes.

Q_M, the photosynthetic energy transformation may be neglected more readily than the two other terms of Eq. (4.11) discussed above. Monteith (1973) showed that Q_M is typically in the range $5–15 \, W \, m^{-2}$ and that it may be neglected, especially when compared with the net radiation.

When the energy budget is not balanced, there will be a change in temperature in the volume considered. From weighting, according to stand density and considering the transpiration of the vegetation, one finds

$$\frac{d\theta}{dt} = \frac{\partial}{\partial x}\left(K_{hx}\frac{\partial \theta}{\partial x}\right) + \frac{\partial}{\partial y}\left(K_{hy}\frac{\partial \theta}{\partial y}\right) + \frac{1}{(H-h)^2}\frac{\partial}{\partial z*}\left(K_{hz}\frac{\partial \theta}{\partial z*}\right)$$

$$+ \frac{1-n_c}{H-h}\frac{1}{c_p\varrho}\frac{\partial R_L}{\partial z*} + \frac{n_c}{H-h}\frac{1}{c_p\varrho}\frac{\partial R_N}{\partial z*} - n_c\frac{L_s}{c_p}P_s. \tag{4.12}$$

In this equation, R_L is the longwave radiation flux over unforested ground, R_N is the net radiation flux within the stand and the last term on the right-hand side takes into account the transpiration of the stand. Transpiration represents a special measure of the plant to protect against overheating of its leaves or needles. Other mechanisms are an increased albedo, which prevents the establishment of extreme temperatures on the leaf surface, and a specific orientation of the leaves in relation to the sun. In this case the area directly exposed to radiation is reduced to a minimum. Flexibility at the point of attachment of the leaves to the branches allows improved convective heat exchange with the ambient air.

However, transpiration is undoubtedly the dominant factor, its importance having been confirmed by various observations. The investigations by Ladefoged (1963) on a beech stand showed that 50–70% of the total energy input is consumed during transpiration. An indication of the huge volumes of water vapor entering the atmosphere through plant transpiration is available in the summary by Larcher (1984). In tropical rainforests it corresponds to about 75% of the total annual precipitation and even in central European mountain forests values of about 40% are reached.

This water is taken up initially through the root system and is then transported from cell to cell and via the specialized ducts to the above-ground parts of the plant. Through the stomata in the outer walls of the epidermal cells, the water vapor eventually leaves the plant and enters the atmosphere.

The opening of the stomata is controlled mainly by a CO_2 cycle and an H_2O cycle (Larcher 1984). The latter mechanism leads to opening of the stomata at high rates of water supply and high ambient temperatures. A lack of water and higher wind speeds cause the stomata to close in order to avoid desiccation of the plant.

The rate of transpiration is controlled by the gradient of the water vapour content between the plant surface and the ambient air (Δs) and by the total diffusion resistance r_i (Hoyningen-Huene 1980):

$$P_s = \frac{\Delta s}{r_i}. \tag{4.13}$$

According to Rutter (1975), the total diffusion resistance is made up of: the resistance of the stomata, r_s, which depends on the specific structure of these openings; the usually rather high cuticular resistance r_k and the boundary layer resistance, r_g, which is controlled mainly by the wind speed.

The factors r_s and r_k act in parallel and thus:

$$\frac{1}{r_{sk}} = \frac{1}{r_s} + \frac{1}{r_k}. \tag{4.14}$$

As the cuticular resistance is mostly rather high, Larcher (1984) reports values of up to $40\,000\,\text{s}\,\text{m}^{-1}$, one may also write $r_{sk} = r_s$.

Depending on the type of tree, r_s may vary within wide limits. For open stomata, a minimum value of $r_s = 400\,\text{s}\,\text{m}^{-1}$ may be considered characteristic for conifers. A review of various measured values of r_s, which may differ by a factor of up to 20, was presented by Jarvis et al. (1975).

The boundary layer resistance, r_g, depends on the wind speed impacting on the leaf and on the structure of the individual leaf. Braden (1982) gave a relation for r_g taking into account the above factors. An approximation may be obtained from: $r_g = 100\,\text{s}\,\text{m}^{-1}$ for $V < 2\,\text{m}\,\text{s}^{-1}$, $r_g = 50\,\text{s}\,\text{m}^{-1}$ for $V > 2\,\text{m}\,\text{s}^{-1}$.

The above values for the individual components of the total diffusion resistance only apply to the individual leaf surfaces. The total resistance of the stand may be calculated with the aid of the leaf area index or, for individual layers, with the leaf area density $b(z)$ as

$$r_c = \frac{r_i}{b} = \frac{r_g + r_s}{b}. \tag{4.15}$$

This eventually leads to the transpiration in Eq. (4.12) through

$$P_s = b \frac{s^* - s}{r_g + r_s}. \tag{4.16}$$

In Eq. (4.16) it is assumed that the air in direct contact with the leaf surface is completely saturated. This allows the specific humidity of the leaf boundary layer to be calculated from the saturation humidity of the ambient air (s^*).

As the opening of stomata is controlled by the various factors outlined above, the transpiration varies widely during the day. Observations of the diurnal variation of the relative humidity in a 5- to 6-m-high young pine plantation (Baumgartner 1956) revealed a notably increased vapor pressure during the daytime.

The moisture reaching the atmosphere through transpiration leads to a pronounced change in the overall humidity field. This has to be taken into account in the budget for the specific humidity:

$$\frac{ds}{dt} = \frac{\partial}{\partial x}\left(K_{hx}\frac{\partial s}{\partial x}\right) + \frac{\partial}{\partial y}\left(K_{hy}\frac{\partial s}{\partial y}\right) + \frac{1}{(H-h)^2}\frac{\partial}{\partial z^*}\left(K_{hz}\frac{\partial s}{\partial z^*}\right) + n_c P_s. \tag{4.17}$$

The terms concerning the longwave radiation flux in Eq. (4.12) are calculated for the unforested part from Brooks (1950)

$$\frac{1}{H-h}\frac{1}{c_p\varrho}\frac{\partial R_L}{\partial z^*} = \frac{1}{H-h}\frac{1}{c_p\varrho}\frac{\partial}{\partial z^*}(R_{L\Downarrow} - R_{L\Uparrow}). \tag{4.18}$$

Here $R_{L\Uparrow}$ describes the radiation flux directed upward from the layer between the reference level and the ground and $R_{L\Downarrow}$ the downward flux from the layer between the upper boundary of the water vapor atmosphere and the reference level. These may be calculated as:

$$R_{L\Downarrow} = \int_z^{z_T} \sigma T^4 \frac{\partial \varepsilon}{\partial z'} \, dz', \tag{4.19}$$

$$R_{L\Uparrow} = \sigma T(h)^4 [1 - \varepsilon(z,h)] + \int_h^z \sigma T^4 \frac{\partial \varepsilon}{\partial z'} \, dz',$$

where z_T is the height to which water vapor is an atmospheric compound, a value of 15 km is used. ε is the emissivity which represents the reemitted fraction of the long-wave radiation entering the water vapor layer. Here, only the emissivity of water vapor is taken into account, while the emissivity of other compounds, such as CO_2, may be neglected.

Measurements by Kuhn (1963) resulted in the following approximations for determining ε:

$$\varepsilon(z, z + \Delta z) = \begin{cases} 0.113 \log(1 + 12.6): & < \log r \leq -4.0 \\ 0.440 + 0.104 \log r: & -4.0 < \log r \leq -3.0 \\ 0.491 + 0.121 \log r: & -3.0 < \log r \leq -1.5 \\ 0.527 + 0.146 \log r: & -1.5 < \log r \leq -1.0 \\ 0.542 + 0.161 \log r: & -1.0 < \log r \leq 0.0 \\ 0.542 + 0.136 \log r: & 0.0 < \log r. \end{cases} \tag{4.20}$$

The absorbing quantity of water vapor r between the heights z and $z + \Delta z$ is calculated from:

$$r(z) = \int_z^{z+\Delta z} \left(\frac{p}{p_0}\right)^{0.85} \varrho s \, dz', \tag{4.21}$$

and has to be entered in Eq. (4.20) in units of $g\,cm^{-2}$.

According to Yamada (1982), the net radiation arriving in the canopy is calculated from the algebraic sum of the shortwave and longwave radiation components:

$$R_N(h_t) = (1 - a_t)S + R_{L\Downarrow}(h_t) - R_{L\Uparrow}(h_t). \tag{4.22}$$

S describes the direct solar radiation and a_t the tree albedo.

Equation (4.19) may be used for the calculation of $R_{L\Downarrow}(h_t)$, whereas $R_{L\Uparrow}(h_t)$ may be calculated from:

$$R_{L\Uparrow}(h_t) = \varepsilon_t \sigma T(h_t)^4 + (1 - \varepsilon_t)R_{L\Downarrow}(h_t). \tag{4.23}$$

The vertical changes in the net radiation in the stand from the crown down to ground level may be approximated rather well from Uchijiama (1961) by an exponential decrease:

$$R_N(z) = R_N(h_t)e^{-k_cL(z)}. \tag{4.24}$$

The extinction coefficient k_c is an empirical constant varying with the time of day (Impens and Lemeur 1969) and type of tree (Larcher 1984). As a typical value $k_c = 0.6$ may be used.

In analogy to Eq. (4.24), in a numerical model the net radiation in the stand is calculated from:

$$R_N(z) = n_c R_N(h_t) \left\{ e^{-k_cL(z)} - n_c \left(1 - \frac{z}{h_t}\right) e^{-k_cL(0)} \right\}. \tag{4.25}$$

The second term in the brackets is used to guarantee that at $n_c = 1$, i.e., when the ground is completely covered with trees, R_N also disappears.

Differentiation of this relation by z results in the divergence of the net radiation flux in Eq. (4.11):

$$\frac{\partial R_N(z)}{\partial z} = n_c R_N(h_t) \left\{ -k_c \frac{\partial L}{\partial z} e^{-k_cL(z)} + \frac{n_c}{h_t} e^{-k_cL(0)} \right\}. \tag{4.26}$$

The surface temperature is calculated from the energy balance equation for the earth's surface under consideration of the vegetation from

$$(1 - n_c)[(1 - a_g)S + R_{L\downarrow}(h) - R_{L\uparrow}(h)] + n_c R_N(h_t)(1 - n_c)e^{-k_c L(0)}$$
$$= Q_H + Q_V + Q_B, \tag{4.27}$$

where a_g is the albedo of the unforested ground.

The shortwave radiation incident on a horizontal area depends on the geographic latitude φ, the time of day and the season. It is calculated from:

$$S = S_0 \cos Z. \tag{4.28}$$

S_0 is the solar constant of 1353 W m^{-2} and Z is the zenith angle. To determine Z, the following relation is applied:

$$\cos Z = \sin \varphi \sin \delta + \cos \varphi \cos \delta \cos t_h. \tag{4.29}$$

Here δ is the angle of declination and t_h the solar hour.

The proportion of the shortwave radiation reaching the ground is only a fraction of that present at the outer boundary of the atmosphere. The attenuation is caused by dispersal and absorption in gases, notably by O_2, O_3 and CO_2. According to Atwater and Brown (1974), this effect may be taken into account by modifying the original version by Kontratyev (1969), by:

$$G = 0.485 + 0.515\left(1.041 - 0.16\sqrt{\frac{9.49\ 10^{-4}\ p_0 + 0.051}{\cos Z}}\right), \tag{4.30}$$

where p_0 is surface pressure in hPa. When the absorption by water vapor (after McDonald 1960)

$$a_w = 0.077\left(\frac{r}{\cos Z}\right)^{0.3} \tag{4.31}$$

is included, we find S at ground level (after Pielke 1984) to be:

$$S = S_0 \cos Z(1 - a_g)(G - a_w). \tag{4.32}$$

The influence of the exposure of an inclined surface on the shortwave radiation is expressed by the geometry factor $\cos i$. i represents the angle between the incident direct solar radiation and the normal to the slope (Fig. 4.2).

According to Kontratyev (1969) there is

$$S_h = S \cos i, \tag{4.33}$$

with

$$\cos i = \cos \alpha \cos Z + \sin \alpha \sin Z \cos(A_s - A_h). \tag{4.34}$$

To calculate the left-hand side of Eq. (4.34), one requires the slope inclination α, zenith angle Z, solar azimuth A_s and slope azimuth A_h. Z may be found from Eq. (4.29), while the other three parameters are found from the following

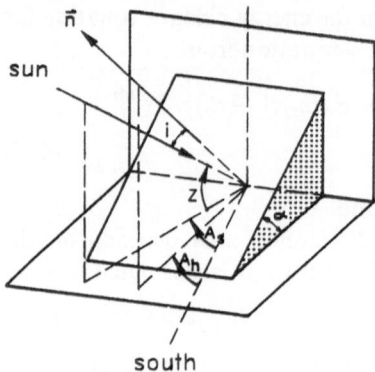

sun

south

Fig. 4.2. Geometric parameters influencing the slope insolation

relations:

$$\alpha = \tan^{-1}\sqrt{\left(\frac{\partial h}{\partial x}\right)^2 + \left(\frac{\partial h}{\partial y}\right)^2}, \tag{4.35}$$

$$A_s = \sin^{-1}\left(\frac{\cos \delta \sin t_h}{\sin Z}\right), \tag{4.36}$$

$$A_h = \tan^{-1}\left(\frac{\partial h}{\partial x}\Big/\frac{\partial h}{\partial y}\right) + \frac{\pi}{2}. \tag{4.37}$$

The net radiation $R_{L\Uparrow}(h)$ from the ground is calculated from:

$$R_{L\Uparrow}(h) = \varepsilon_g \sigma T(h)^4 + (1 - \varepsilon_g) R_{L\Downarrow}(h), \tag{4.38}$$

where ε_g is the ground emissivity.

The turbulent sensible heat flux Q_H, the turbulent latent heat flux Q_V and the soil heat flux Q_B are calculated from:

$$Q_V = \varrho L_s u_* s_*, \tag{4.39}$$

$$Q_H = \varrho c_p u_* \theta_*, \tag{4.40}$$

$$Q_B = -\lambda_s \frac{\partial T_s}{\partial z_s}. \tag{4.41}$$

In these equations, u_* is the friction velocity, θ_* is the temperature scale and s_* is the water vapor scale. T_s is the temperature in the ground, λ_s is the heat conductivity and z_s the depth within the ground (positive downwards). T_s is found by solving the equation for heat conduction in the soil using as boundary conditions a constant temperature at a depth of 50 cm and the surface temperature calculated from Eq. (4.27).

With the aid of a Newton–Raphson iteration, $T(h)$ is calculated diagnostically from the energy flux budget. If the sum of the energy fluxes in Eq. (4.27) is referred to as Q, there is:

$$T(h)^m = T(h)^{m-1} - \frac{Q[T(h)]^{m-1}}{Q'[T(h)]^{m-1}},$$

(4.42)

with $Q'[T(h)] = \partial Q / \partial T(h)$. The iteration is stopped when the difference of Q between two following iterations m and $m - 1$ is smaller than 10^{-5}.

The actual ground temperature thus determined leads to the lower boundary condition for θ through the definition of the potential temperature. For u, v, w, p' and E at ground level, the boundary conditions described in Section 3.1.2 are used. As the specific moisture appears as a variable in the extended version of the model, boundary conditions have also to be set for this variable. We use:

$$s(h) = s(z_1) + c_f[s^*(h) - s(z_1)],$$

(4.43)

with $s^*(h)$ as the saturation humidity at ground level and $s(z_1)$ as the specific humidity of the atmosphere close to the ground. For practical purposes, the value of s in the first level of the numerical grid is used for this. c_f is the relation between the moisture content of the soil, W_G, and the field capacity, W_{FC}, which represents the highest moisture content possible. To determine W_G, one solves:

$$\frac{\partial W_G}{\partial t} = \frac{1}{\varrho_w}\left[\alpha^*(W_G - W_{FC}) - \frac{Q_v}{L_s}\right],$$

(4.44)

where ϱ_w is the density of water and α^* is the soil capillarity which depends on the specific site, as does W_{FC}. Through the root system, any vegetation covering the ground will withdraw moisture from the soil and this then enters the atmosphere by transpiration. McCumber (1980) describes a method for taking this effect into account. It requires a knowledge of the vertical root distribution and the depth of root penetration into the ground. However, such information is usually not known.

For $n_c = 0$ and $n_c = 1$ characteristic values of field capacity, soil capillarity and heat conductivity are used for uncovered ground and for ground completely covered by vegetation, respectively. For intermediate cases these parameters are weighted with the fraction of the area covered with trees, according to Eq. (4.1).

All other boundary conditions and the numerical solution of the equations are rather similar to those used in the simplified version of the model (Sect. 3.1). Therefore, only the differences between the two model versions will be outlined below.

Just as the potential temperature is specified by a vertical profile of the basic state, so is the specific humidity. At the upper boundary a characteristic value is given for the air mass, in which the mesoscale model domain is embedded. This value remains constant with time. At the outflow boundaries the normal derivatives are assumed to disappear.

In complex terrain the potential temperature at the ground must be considered not as constant at the initial time, but as adjusted according to the elevation. For this purpose, $\theta(h)$ is considered equal to the potential temperature of the basic state at the appropriate height of the lower boundary.

For integration with time, the forward-directed Euler scheme is used and the advection terms are approximated upstream.

4.2 Investigations in a Horizontally Homogeneous Stand

4.2.1 Determination of the Optimum Drag Coefficient

In the extended version of the model vegetation is taken into account through an additional friction term in the respective equation of motion. This term results from the retardation of the wind through exchange of momentum on the stand elements.

In analogy to the conclusions found from the flow around bodies of simple geometry (slab, cylinder), the resistance of a highly structured and porous stand is approximated by:

$$F = c_d \varrho b(z) u^2. \tag{4.45}$$

Results from the investigation of flow around groups of trees offer the possibility of calculating an optimum drag coefficient. For this purpose, a mean vertical profile of wind speed $\hat{u}$ is calculated from the simulation results shown in Fig. 3.36a–e.

Furthermore, it is possible to calculate the forest density n_c as the relation between the projection of all tree sections on the ground and the total model area. This results in a value of $n_c = 0.15$.

When the same meteorological input parameters as are used in a one-dimensional version of the model are used in the three-dimensional simulation, a drag coefficient can be found which results in a best-fit vertical profile. The results of one-dimensional calculations with different drag coefficients are summarized in Table 4.1. The table presents the mean deviations (in cm s^{-1}) of the one-dimensional wind profile from the mean vertical profile $\hat{u}$ of the three-dimensional simulation. c_{d_0} fluctuates within wide limits from 0.05 to 2.0. Nevertheless, the mean differences are fairly large with the smallest value of 12 cm s^{-1} at $c_{d_0} = 0.1$. One has thus to expect some control of c_d by the forest

Table 4.1 Mean wind speed deviation (m s^{-1}) between the results derived from one-dimensional simulations and from three-dimensional simulations (examples from Fig. 4.3b are in *italics*)

c_{d0}	0.05	0.10	0.20	0.40	0.60	0.80	1.00	2.00
$c_d = c_{d0} \cdot n_c^2$	*0.09*	0.05	*0.05*	0.06	0.08	0.10	0.13	0.18
$c_d = c_{d0} \cdot n_c$	0.10	0.08	0.06	0.07	0.13	0.18	0.19	0.22
$c_d = c_{d0}$	0.13	0.12	0.14	0.17	0.20	0.22	0.23	*0.27*

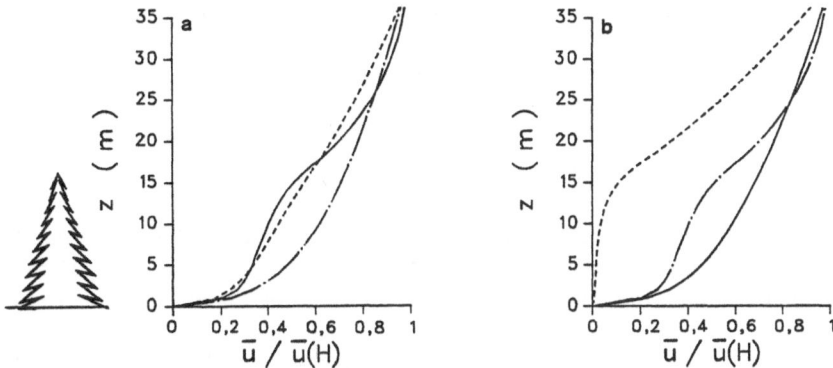

Fig. 4.3. Vertical wind profiles. **a** *Solid line* 3-D (averaged); *dashed line* 1-D ($c_d = 0.2n_c^2$), *dashed* and *dotted line* 1-D ($c_d = 0$). **b** *Solid line* c_{do} n_c^2 ($c_{do} = 0.05$); *dashed* and *dotted line* c_{do} n_c^2 ($c_{do} = 0.2$); *dashed line* c_{do} ($c_{do} = 2$)

density and this leads to a notable improvement. The smallest deviation of 5.2 cm s^{-1} is simulated for $c_d = c_{do} \cdot n_c^2$ with $c_{do} = 0.2$.

The normalized vertical wind profile corresponding to this combination of parameters is presented in Fig. 4.3a together with $\hat{u}/u(H)$ and the log-profile (without stand).

The evaluation of the other three-dimensional simulations of flow around isolated trees leads to approximately similar results. The minimum of the mean deviation is always in the range $c_{do} = 0.1\text{--}0.4$ when multiplied with n_c or n_c^2.

Figure 4.3b shows the vertical profiles calculated for the cases given in italics in Table 4.1. For a large drag coefficient the shape of the profile of $\hat{u}$ is reproduced rather well. The wind speed close to the ground, however, is reduced so drastically that the deviation from the profile of the three-dimensional simulation becomes very large indeed. The canopy friction terms is almost without effect when the value chosen for c_d is too small. In this case there is no notable influence of the vegetation on the wind pattern.

4.2.2 Influence of Stand Parameters on Temperature, Humidity, Wind and Turbulence

The microclimate in a forest differs from that of an area covered only by low vegetation. However, despite notable quantitative and qualitative differences, it is not always possible to arrive at generally applicable conclusions because of the multitude of controlling parameters. At around midday in a deciduous forest stratification is usually stable, whereas, prior to the appearance of the leaves in the same stand, unstable thermal conditions develop (Chroust 1968). The observations at any given site depend on the season, time of day and actual weather conditions. Great care has to be taken when comparing measurements from different localities. Differences may be caused by, for example: orography (mountains or plains, north- or south-facing slopes), weather conditions (air mass, regional wind speeds, supply of radiation), soil conditions (type of soil, water supply), geometry of the individual trees (height, diameter and density of the crown, height of the trunk) and structure of the stand (arrangement of trees, forest density).

In order to isolate the influence of the different stand parameters on the distribution of meteorological variables like wind, humidity and temperature, a number of simulations were carried out with the one-dimensional version of the model.

An 18-m-high coniferous forest with a leaf area index of 7 $m^2\,m^{-2}$ is selected as the model stand. The highest leaf area density of $b_{max} = 0.7\,m^2\,m^{-3}$ is encountered at about 6-m height, and above and below this an almost linear decrease is assumed (Hicks et al. 1975). The other stand-specific parameters were selected as follows: an extinction coefficient of 0.6, an albedo of 0.3, an emissivity of 0.98 and a drag coefficient of 0.2 (multiplied by the square of the forest density).

As meteorological parameters the components of the geostrophic wind $(u_g = 5\,m\,s^{-1},\quad v_g = 0)$, a large-scale temperature profile $(\theta_0 = 290\,K,\ \partial\tilde{\theta}/\partial z = 0.35\,K/100\,m)$ and a large-scale humidity profile (relative humidity of 30% = constant) were assumed.

Other local parameters are a geographic latitude of 48°, a roughness length of 0.05 m, a heat conductivity of the soil with 1.25 $W\,m^{-1}\,K^{-1}$, a temperature of 290 K at the depth of 50 cm, an albedo of 0.1 and an emissivity of 0.98 of the unforested area.

Mean wind speed, temperature, humidity and turbulence fluctuate considerably close to the ground and above the canopy, whereas changes at greater heights are only small. In order to resolve properly these variations, the grid levels are not arranged at a uniform distance but, instead, a larger number are arranged in the lowermost part of the atmosphere. The elevations of the 25 levels above ground are: 0, 3, 6, 9, 12, 15, 18, 21, 30, 50, 80, 120, 200, 400, 600, 800, 1000, 1200, 1400, 1800, 2200, 2600, 3000, 3500 and 4000 m. To solve the heat conduction equation in the soil, four additional levels are selected at depths of 2, 10, 25 and 50 cm.

As a result of the good resolution near the ground and of the unstable stratification prevalent at around midday, together with a large diffusion coefficient, a time step of 1 s was used for the calculations at this time.

The horizontally homogeneous stand assumed in these calculations will be encountered only rarely in nature as, often, the top of the canopy is protruded by isolated taller trees. It is thus not surprising to observe differences of up to 1°C in mean air temperature at crown height over a distance of 30 m in a relatively homogeneous stand (Bergen 1971). Such differences lead to local wind speeds of about 0.1 $m\,s^{-1}$ and these make the assumption of horizontal homogeneity scarcely realistic.

Furthermore, in a one-dimensional model mean vertical velocities are not possible. The cold air forming at night at crown height will migrate downward under the influence of gravity. This natural vertical advection is excluded from the model used here.

Consequently, the results outlined here apply to a very large forest area with minimal local differences. The results may be compared only qualitatively with actual measurements as field measurements refer to local conditions and it is not possible to derive from them general conclusions averaged over space.

In the following discussions the conditions of radiation prevalent on a day in July (21st July) will be used as a basis. The integration with time starts at midnight and is carried out for 4 days of real-time. Already, after the second day, there are notable diurnal fluctuations which are similar to one another but are not exactly identical. The results are shown below for 0000-2400 LST of the fourth day.

First the influence of different forest densities is studied. The results for a very dense forest ($n_c = 1$), an open stand ($n_c = 0.5$) and unforested open ground ($n_c = 0$) will be compared. The other input parameters described above will remain unchanged. The comparison of the results for $n_c > 0$ and $n_c = 0$ is of particular interest because it facilitates an estimate of the climatic influences exerted by large-scale deforestation or by progressive forest decline.

At $n_c = 1$, the dense canopy shields the ground from direct insolation. The soil temperature is only controlled by Q_H, Q_V and Q_B, showing a low amplitude diurnal fluctuation of only 4°C. In contrast to this, the interface between the upper part of the forest and the atmosphere above represents an energetically important surface along which pronounced diurnal variations of the various components of the radiation budget and of temperature may be observed (Jarvis et al. 1975; Mayer 1982; Heisler 1985). In Fig. 4.4a the diurnal variation of the potential temperature is shown for the simulation of a dense forest. During the night (0200–0600 LST) an unstable stratification is found inside the stand itself. This results from the rapid cooling of air due to the longwave radiation near the top of the canopy. Above this, a stable stratified atmosphere is simulated. During the daytime the canopy heats up rather rapidly, especially in those regions where the heating function $\partial R_{NP}/\partial z$ attains a maximum. The position of this maximum depends on the value selected for the extinction coefficient in the exponential law of radiation in the stand and on the vertical distribution of the stand elements.

According to Eq. (4.25) only a small fraction of the direct solar radiation reaches the ground, and a stable stratification is established between ground and crown and is maintained throughout the day. The temperature difference between the elevated maximum and the ground is about 4°C around midday.

The simulated diurnal variation of θ in the stand agrees qualitatively with observed field data. For an open pine stand, Göhre and Lützke (1956) report a temperature difference of about 2°C between ground and crown, whereas Baumgartner (1956) observed a difference of 6°C in a thinned-out pine stand. Differences of up to 8°C were reported by Hosker et al. (1974) , also in a pine

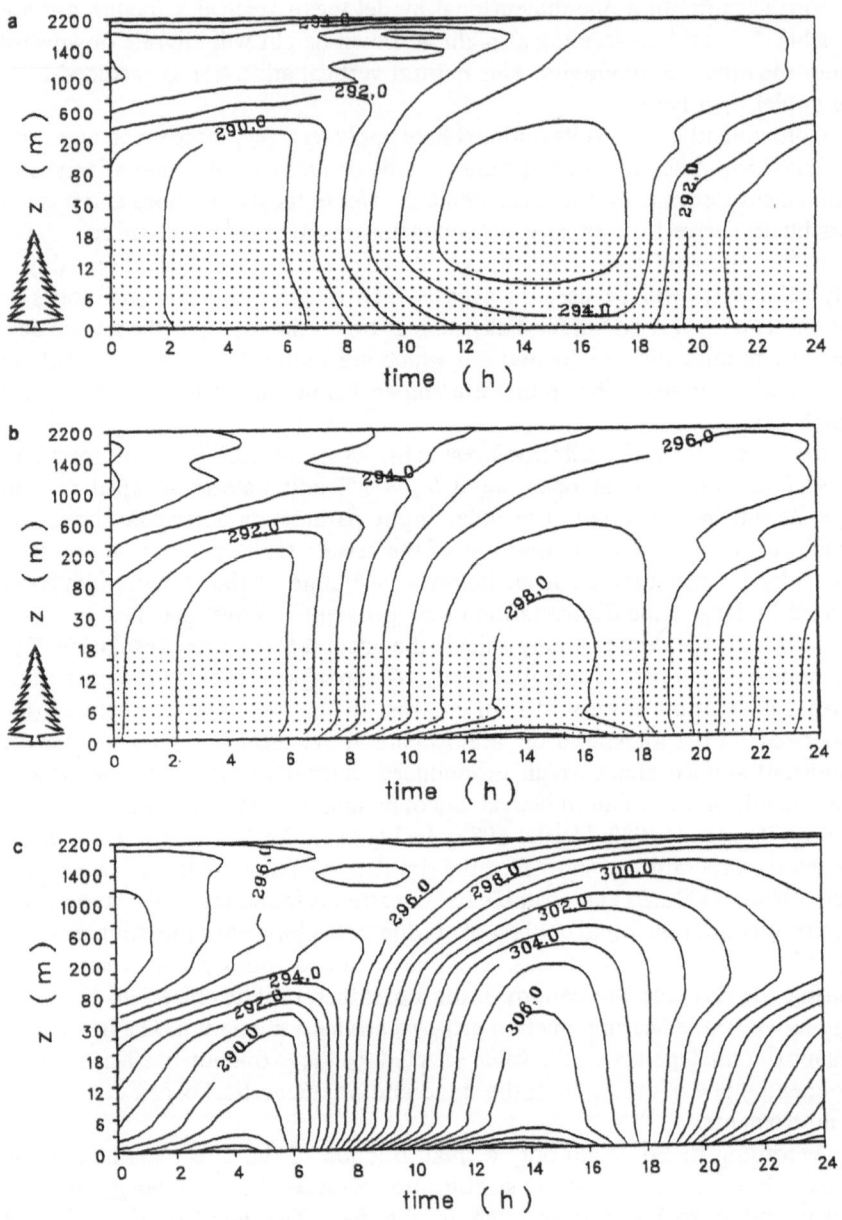

Fig. 4.4. Diurnal variation of potential temperature (K, intervals 2 K). **a** $n_c = 1$; **b** $n_c = 0.5$; **c** $n_c = 0$. The *dotted area* indicates the canopy

stand. As pointed out above, these values are not strictly comparable due to the multitude of site parameters influencing the temperature distribution.

A reduction of the forest density to $n_c = 0.5$ leads to a complete change in the structure of the diurnal variation of the potential temperature. Under this condition, the nocturnal cooling and the warming at around midday take place along the ground as well as higher in the stand. This leads to a stable stratified atmosphere near ground level from sunset to sunrise and to an unstable boundary layer during the daytime. With an amplitude of 10°C, the diurnal variation of θ_0 is double the value encountered in the simulation with $n_c = 1$.

Simulations with forest densities between 0.6 and 0.9 show that the midday change from stable stratification between ground and crown to unstable stratification occurs at a value of $n_c = 0.8$. In this case, one finds a temperature decrease with height close to the ground, then an increase up to the crown and, above that, a decrease again. This illustrates the pronounced influence exerted by the forest density on the results of a simulation.

In the case of unforested open ground ($n_c = 0$) an unstable stratified atmosphere is developed during the daytime, the diurnal variation of which extends to a great height (Fig. 4.4c). The cooling near the ground leads to a stable stratification throughout the night. As a result of the unimpeded insolation the net radiation reaches values of about 800 W m^{-2} (Fig. 4.5), leading to a temperature increase to about 35°C at a height of 3 m at 1400 LST. There is a pronounced diurnal variation with an amplitude of 27°C. These results agree well with the observations of, e.g., Clarke et al. (1971).

When comparing the profiles for 0000 and 2400 LST, an almost stationary situation is found for the lower portion of the atmosphere, but not in the higher parts between 500 and 1000 m where the air is heated up by about 2 K day^{-1}. The warming of the air rising to this height by diffusion cannot be compensated by radiative cooling. This almost constant warming effect will either not be

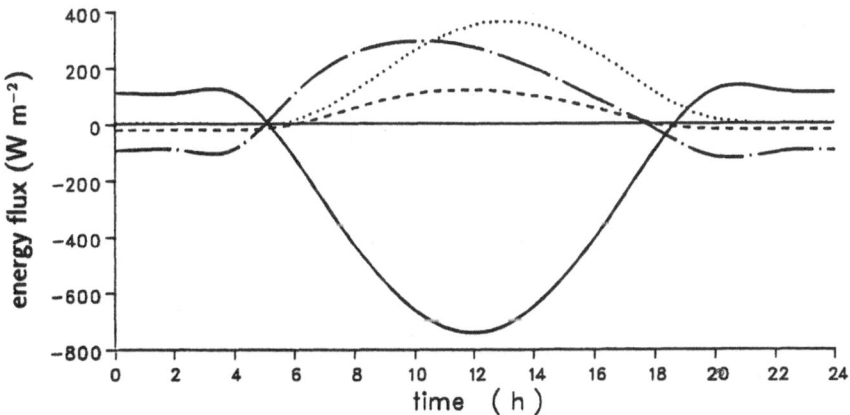

Fig. 4.5. Diurnal energy flux variation on the ground for $n_c = 0$. *Solid line* R_N; *dashed* and *dotted line* Q_B; *dashed line* Q_H; *dotted line* Q_V

developed, or only be developed to an attenuated degree when advective processes in horizontally inhomogeneous situations are taken into account.

The distribution of relative humidity is closely tied to the temperature field. Measurements by Baumgartner (1956) in a dense, 5- to 6-m-high pine stand showed humidities of up to 95% in the early morning hours. Close to the ground this maximum is maintained throughout the day because of the low degree of mixing of the moist air in the stand with the dry air surrounding it. A minimum of 40% is encountered around midday within the canopy.

The same relative humidity pattern is observed in the simulations (Fig. 4.6). Low temperatures are correlated with high relative humidities and vice versa. The distribution of the specific humidity is essentially controlled by the sources, i.e. the soil and the evaporation in the stand. As a result, the maximum is found near the ground at night and in the canopy during the daytime.

Comparative humidity measurements in forests and over open ground were carried out by Burger (1951) and Lützke (1967). Their data sets both show a pronounced diurnal variation, the relative humidity in the forest being at all times higher than over the surrounding meadows.

The results of the simulations for $n_c = 1$ and $n_c = 0$ (Fig. 4.7) again show a qualitatively satisfactory agreement with the observed situations.

From the temperature and humidity fields, the Bowen ratio (Bo) may be calculated. It represents the relation between the turbulent sensible heat flux and the turbulent latent heat flux and reads as $Bo = Q_H/Q_V$. A smaller value of Bo implies that less energy is available to warm the air as the heat is consumed by evaporative processes.

This ratio is frequently used in practical applications and, as a result, a number of measurements are available. Typical values of Bo in the forest range from 0.2 to 0.8 (Larcher 1984). When, in addition, the season and time of day are considered, Bo varies within rather wide limits of −1 to 10 (Steward and Thom

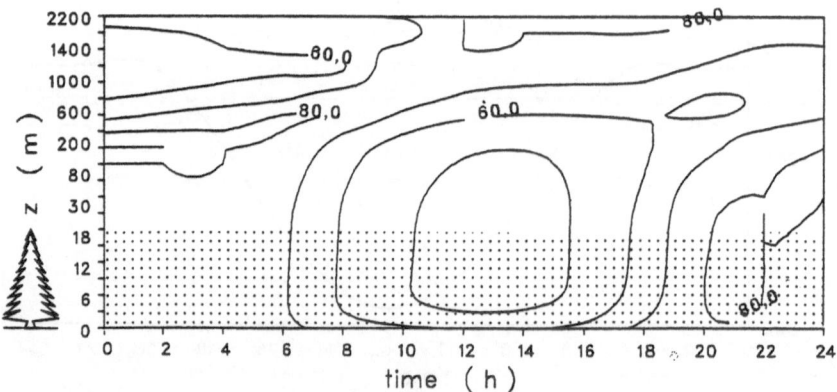

Fig. 4.6. Diurnal variation of relative humidity at $n_c = 1$ (%, intervals 10%). The *dotted area* indicates the canopy

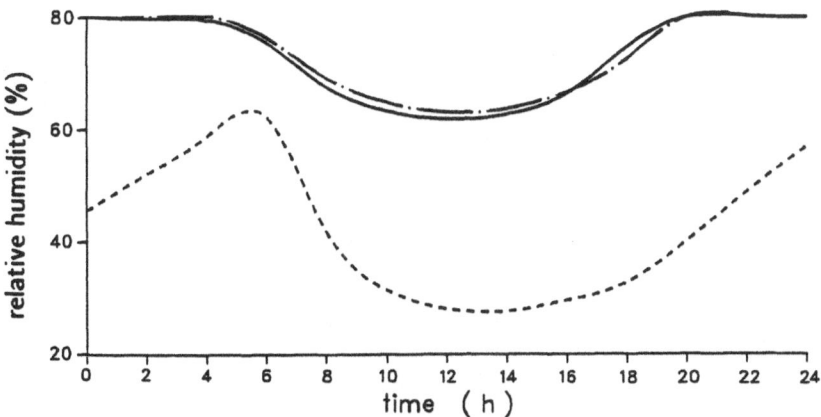

Fig. 4.7. Diurnal variation of relative humidity at 3 m height. *Solid line* $n_c = 1$, *dashed and dotted line* $n_c = 0.5$, *dashed line* $n_c = 0$

Table 4.2. Calculated diurnal variation of Bowen ratios for open ground and different stand densities

Time (h)	$n_c = 0.0$	$n_c = 0.5$	$n_c = 1.0$
0	− 3.4	− 3.3	1.0
2	− 5.1	0.0	0.9
4	− 20.7	0.8	0.8
6	− 28.4	1.4	0.5
8	0.5	1.1	− 18.1
10	0.4	1.1	− 5.2
12	0.3	1.0	− 3.3
14	0.2	1.0	− 2.4
16	0.2	1.0	− 1.6
18	0.0	1.1	− 1.1
20	− 1.3	0.1	0.7
22	− 2.6	− 4.1	0.8
24	− 3.2	− 3.0	0.9

1973) or −0.8 to 2.4 (Munro 1985). The Bowen ratios obtained from the simulations presented here are given in Table 4.2. The high variability known from field data is also evident here.

The vertical transport of heat and moisture in the one-dimensional model takes place by diffusion. Its intensity is described by the turbulent diffusion coefficient which is controlled notably by the vertical shear of the horizontal wind. A prominent characteristic of the climate within a stand is the very low wind speed with small diurnal variation and this was also observed by, e.g., Baumgartner (1956), Bergen (1971) and Oliver (1971).

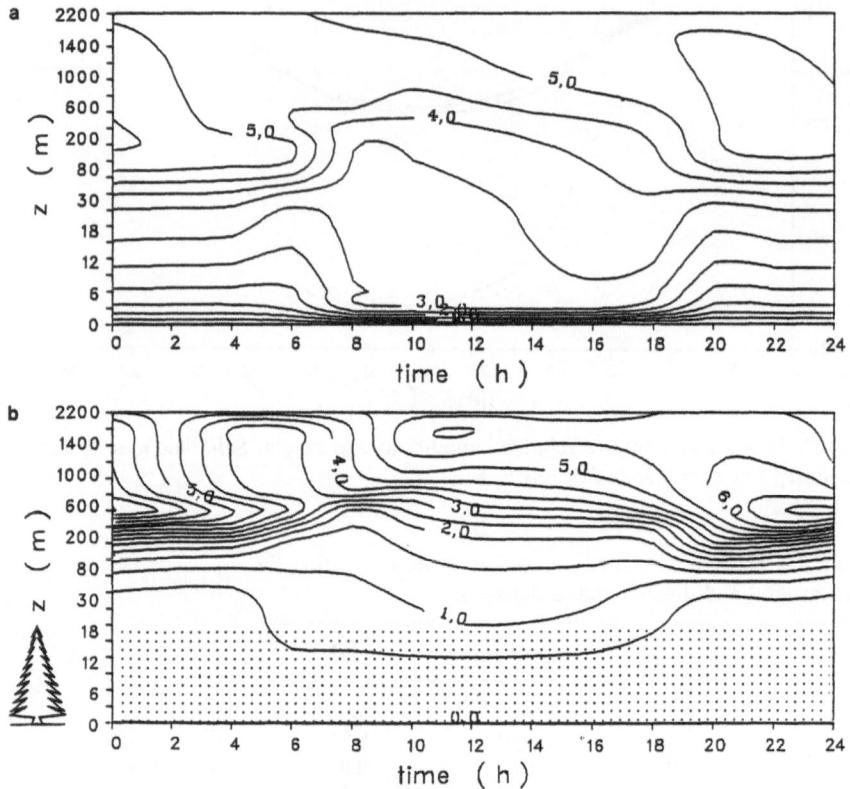

Fig. 4.8. Diurnal variation of wind speed for **a** $n_c = 0$; **b** $n_c = 1$ (m s^{-1}, intervals 0.5 m s^{-1}). The *dotted area* indicates the canopy

Above this calm zone among the trunks, a strong upward increase in wind speed is observed in the upper part of the canopy. A number of observations show a secondary maximum between the ground and the lower boundary of the canopy. From the literature, Kurata (1982) found that this phenomenon is only developed during unstable stratification.

In Fig. 4.8a the simulated diurnal variation of the wind speed over open ground is presented. Close to the ground larger values are developed during the daytime than during the night. This can be ascribed to an increased mixing of the atmosphere which leads to a stronger coupling of wind close to the ground with the geostrophic wind. Furthermore, a zone of high nocturnal wind speeds is simulated for a height of 200 m. At this level, V is up to 20% higher than the value of the geostrophic wind. This boundary layer phenomenon is known in the literature as the *low-level jet* (LLJ). It only occurs in rotating systems, where, however, it represents a permanent feature (e.g., Wippermann 1973). The intensity and position of this maximum are controlled mainly by the thermal

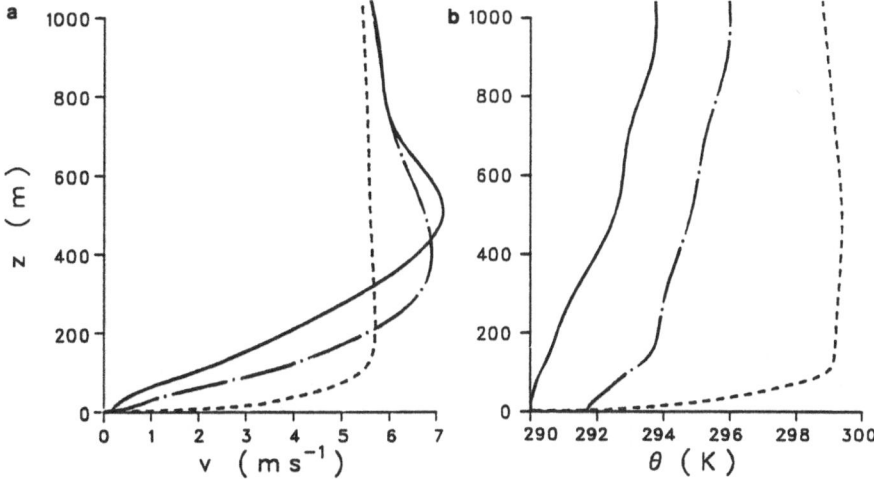

Fig. 4.9. Vertical profiles of **a** wind speed and **b** potential temperature at 2400 LST. *Solid line $n_c = 1$; dashed and dotted line $n_c = 0.5$; dashed line $n_c = 0$*

stratification. This becomes evident from Fig. 4.9 in which vertical profiles of V and θ at 2400 LST are given for $n_c = 0, 0.5$ and 1. The three simulations show a pronounced stable layer which extends upward for 200–800 m. Above this, a neutral, or even slightly unstable, stratification is developed which results from the temperature surplus of the preceding day. The large temperature difference between the three simulations found at a height of 1000 m is caused by the different rates of heating in the zone close to the ground during the midday hours. This large heat surplus is dissipated in the boundary layer through diffusion, and is not compensated completely by radiative cooling.

For the simulation with a dense stand (Fig. 4.8b), the calculated ground-level wind speed is much smaller than over unforested ground, whereas the maximum of the low-level jet is much higher. The reduction of V in the stand can be ascribed to increased friction exerted by the branches and leaves.

The distribution of the mean variables discussed so far is determined mostly by field experiments on the situation within canopies. Considerably fewer data are available for the turbulence parameters. However, knowledge of these could markedly increase our understanding of the role which tall trees in the control of air flow in the lowermost levels of the atmosphere.

Over open ground a maximum of the turbulent kinetic energy is usually observed near the ground level and above that there is a decrease with height. On the other hand, for the situation within the canopy, a number of authors (Raynor 1971; Wilson and Shaw 1977; Bradley and Mulhearn 1983) observed an increase with height from ground level and a maximum within the canopy. McBean (1968) found the turbulence intensity in a stand to be larger than, or at least equal to, that over open ground.

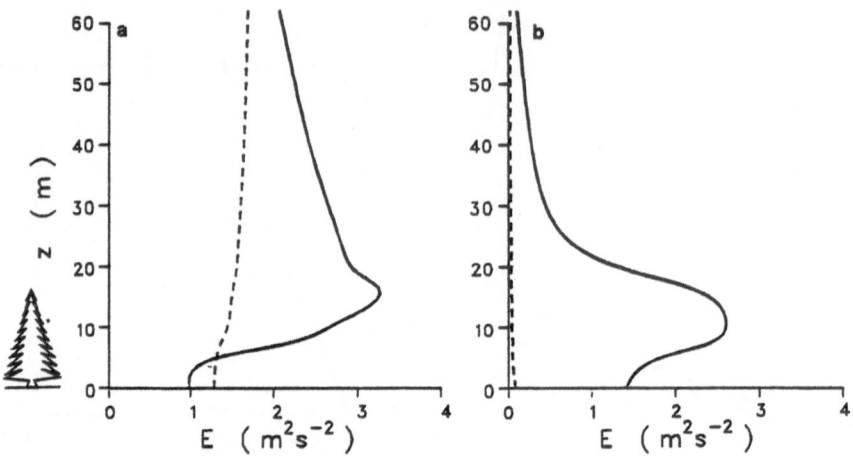

Fig. 4.10. Vertical profiles of turbulent kinetic energy at **a** 1200 LST and **b** 2400 LST. *Solid line $n_c = 1$; dashed line $n_c = 0$*

In Fig. 4.10a vertical profiles of the turbulent kinetic energy at 1200 LST are presented for $n_c = 0$ and $n_c = 1$. In the forest, the maximum of E is twice that over open ground. The maximum is located in the upper part of the canopy. Turbulence here is caused mainly by the production of strong wind shear. The unstable stratification above the stand increases the turbulence through the conversion of potential energy into turbulent kinetic energy. Over the entire tree height, E is larger for $n_c = 1$ than for $n_c = 0$. This can be ascribed to a stronger retardation of the wind in the canopy, leading to a conversion of kinetic energy of the mean flow into turbulent kinetic energy. When E is dissipated, this causes an increase in internal energy.

Over the unforested plain, E is smaller by almost two orders of magnitude during the night than during the day. Production from wind shear and dissipation in the intensely stable stratified atmosphere permit only small values of E. At the same time, within the stand itself unstable stratification is simulated and it it thus not surprising that a large turbulence is found in the forest, especially during the night (Fig. 4.10b).

The turbulent diffusion coefficient K_m is calculated from the turbulence energy with the aid of the mixing length and empirical profile functions. However, it can also be derived from field observations. Uchijiama (1962) and Denmead (1964) noted an increase of K_m from the ground up to the crown, while Landsberg and Thom (1971) calculated a near-constant diffusion coefficient over wide portions of the stand.

The diurnal variations of K_m over forested and unforested ground are shown in Fig. 4.11. Although the turbulent kinetic energy values in these two cases differ notably, the patterns of K_m above the stand show great similarity. Only the absolute value for $n_c = 1$ is larger than for $n_c = 0$. During the night, however,

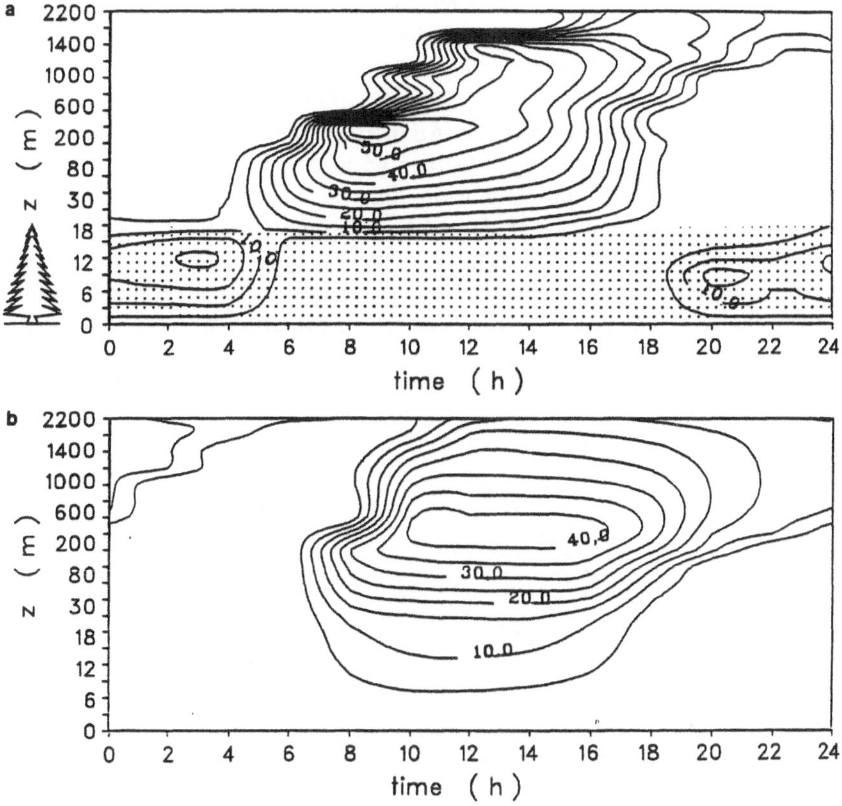

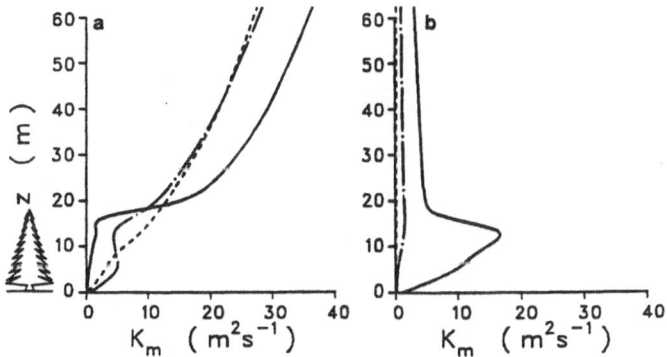

Fig. 4.11. Diurnal variation of turbulent diffusion coefficient for **a** $n_c = 1$ and **b** $n_c = 0$ ($m^2 s^{-1}$, intervals $5 m^2 s^{-1}$). The *dotted area* indicates the canopy

Fig. 4.12. Vertical profiles of turbulent diffusion coefficient at **a** 1200 LST and **b** 2400 LST. *Solid line* $n_c = 1$; *dashed and dotted line* $n_c = 0.5$, *dashed line* $n_c = 0$

totally different conditions are simulated for the lowermost 20 m of the boundary layer. This again can be ascribed to the unstable stratification then forming and to the resulting large values of K_m.

The vertical profiles of K_m at 1200 and 2400 LST (Fig. 4.12) support the structures expected from observations. An increase is observed within the stand, depending on the forest density, together with an almost constant diffusion coefficient. This leads to the conclusion that the observed differences in the vertical profiles of K_m may be explained by differences in stand structure.

In the simulations discussed so far, a vertical profile of the leaf area density which is characteristic of a coniferous forest was assumed for the stand. To study the influence of a different leaf area density, a simulation was carried out for

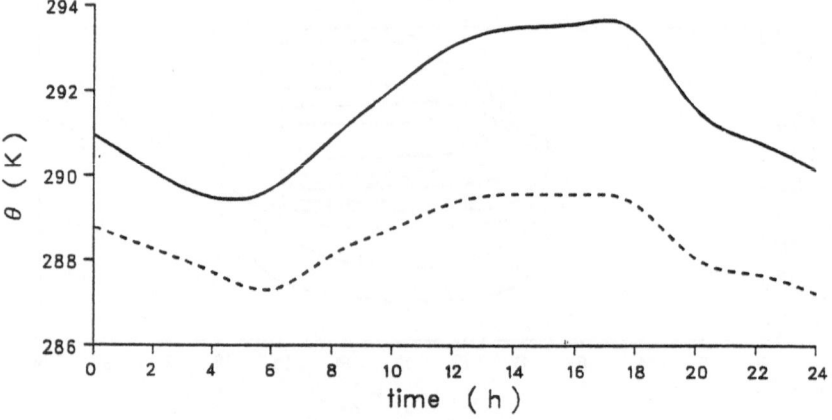

Fig. 4.13. Diurnal variation of surface temperature in coniferous forest (*solid line*) and deciduous forest (*dashed line*) for a tree height of 18 m

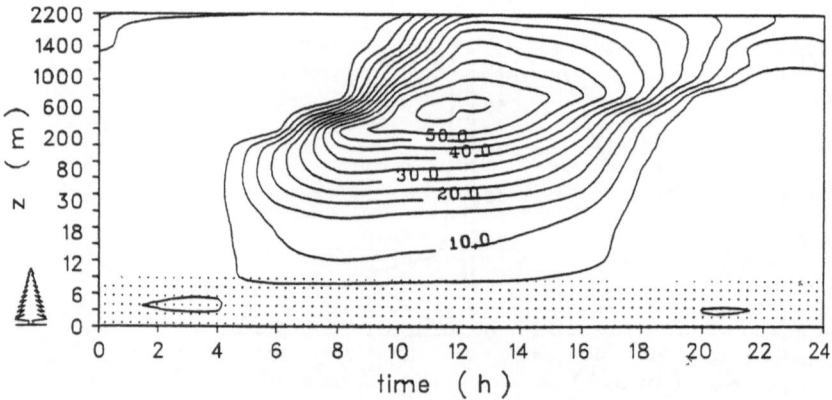

Fig. 4.14. Diurnal variation of turbulent diffusion coefficient for a 9-m-high conifer stand with a density of $n_c = 1$ ($m^2 \, s^{-1}$, intervals 5 $m^2 \, s^{-1}$). The *dotted area* indicates the canopy

Table 4.3. Extreme values of different meteorological variables

Type of tree	h_t	n_c	$\theta_{0\,max}$	$\theta_{0\,min}$	u_{max}	$K_{mz,\,max}$
Conifer	18	1.0	293.5	289.6	7.14	60.1
Conifer	18	0.5	300.6	290.1	6.75	61.7
		0.0	309.8	286.1	5.96	45.8
Conifer	9	1.0	296.7	291.8	6.64	57.8
Deciduous	18	1.0	289.8	287.2	6.96	58.6

$n_c = 1$ and $b(z)$ for a deciduous stand. The leaf area density profile selected shows a pronounced maximum of $1.4\ m^2\ m^{-3}$ in the upper part of the canopy, with very low values in the lower 6 m, i.e., the trunk zone (Rauner 1976).

The increased leaf area density in the uppper part of the canopy leads to a pronounced decrease in the net radiation in the stand. This results in a lower amplitude of the diurnal ground temperature variations of only 3°C (Fig. 4.13). The patterns of the other meteorological variables calculated show only small quantitative differences.

In another simulation for a coniferous stand, the tree height was reduced from 18 to 9 m. In this case, the diffusion coefficient during the night shows the most pronounced difference from the previous simulations. Although the maximum in the canopy is maintained, its value is reduced by a factor of 3 (Fig. 4.14). In Table 4.3 the extreme values of different variables are summarized for the various one-dimensional simulations performed.

4.3 Investigations in a Horizontally Inhomogeneous Stand

4.3.1 The Influence of Deforestation Around the Western Runway of Frankfurt Airport

On April 17th 1984, the operating company, FAG, of the Frankfurt International Airport opened a new north–south runway which, at its northern end, connects with the western end of the existing east-west runway.

For this runway existing forest areas had to be cut down. International aviation regulations demand a lateral safety strip of 300 m on either side of the runway and the German Federal Department of Transportation requires a 900-m-long strip free of obstacles ahead of the end of the runway. This resulted in a 600-m-wide and 3.2-km-long clearing being cut into the existing forest.

Figure 4.15 shows the situation prior to and after construction of the runway. Parallel to and to the west of the clearing there is an additional strip, along which four overhead powerlines are routed. These had to be relocated to bypass the runway farther south, necessitating further deforestation.

According to FAG a total of about 500 ha forest had to be cleared, one-quarter of which will be reafforested. In view of this drastic local modification of

a   b

Fig. 4.15. Aerial photographs of the Frankfurt International Airport **a** prior to and **b** after construction of the western runway

the landscape, significant repercussions on the local climate had to be expected. Consequently, DWD (German weather service) prepared a climatological study (Terpitz 1981), to supplement the earlier studies of 1965, 1966 and 1967. In this assessment a distinction was made between influences over the clearing itself and areas further away from the runway.

It was found that the conditions would change, particularly over the planned clearing, as the previous forest climate would become an open ground climate. This would result in a larger amplitude of the diurnal temperature variation with much higher values around midday and a stronger cooling during the night. As relative humidity is closely tied to temperature, during the cooler morning hours the probability of near surface fog formation will be more frequent than hitherto. Wind speed in the clearing will increase as the greater roughness caused by the trees has been eliminated and local wind systems (forest winds) may develop.

The climatological study came to the conclusion that, especially during conditions of comparatively calm radiation, a separate local climate may develop over the deforested areas. Furthermore, it was concluded 'that far-reaching influences on the surrounding villages can be largely excluded', i.e., no influences on the regional climate were to be expected.

The essentially exploratory character of the DWD assessment was ascribed in its concluding remarks to the fact that only partial data were available. It is in such cases that calculations with numerical simulation models are able to provide constructive information for the decision makers concerned.

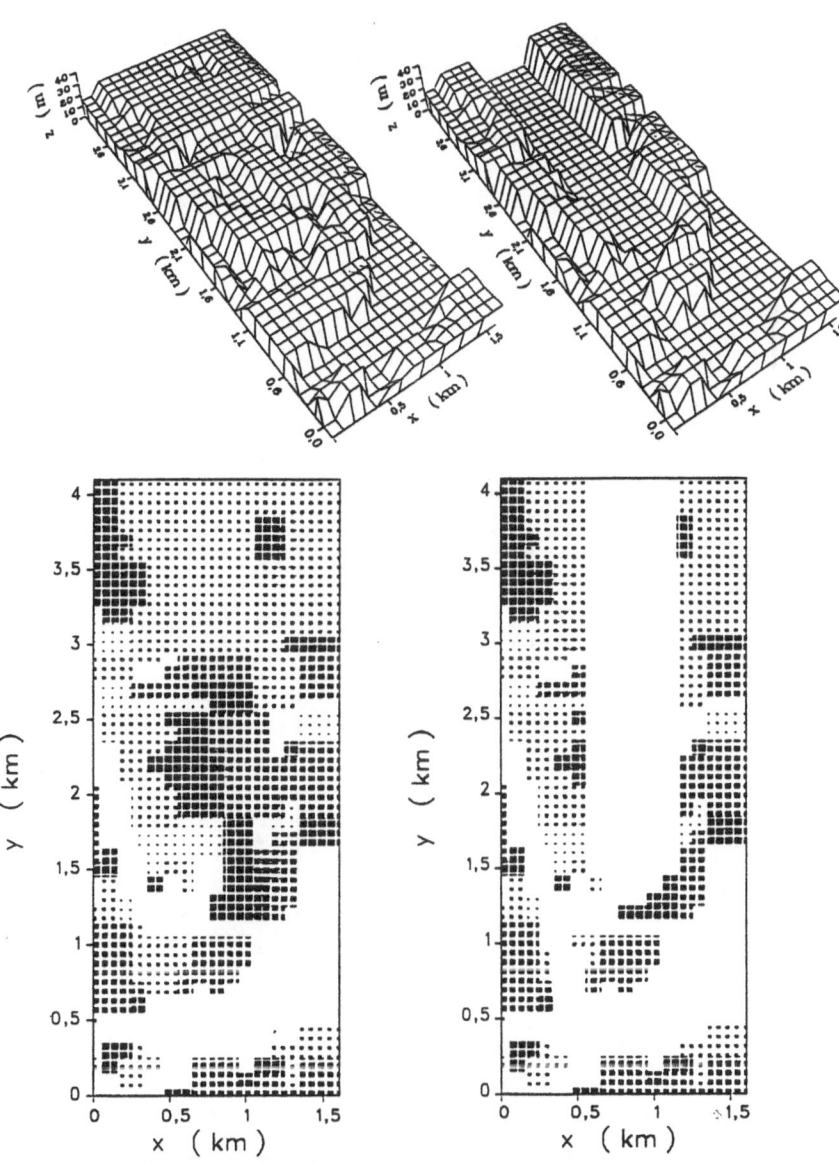

Fig. 4.16. Distribution of tree height h_t (*above*) and stand density n_c (*below*) prior to deforestation (*left*) and thereafter (*right*). Intensity of shading is proportional to n_c.

To illustrate this, the results of two simulations, representing the situation prior to and after deforestation, are presented. By comparing the results, conclusions may be drawn as to the potential changes in local climate.

Differences in orography in the vicinity of the western runway are low and thus may be neglected. The most important local parameter is the forest itself, i.e. the areal distribution of types and heights of trees and stand density. From aerial photographs these parameters were obtained as representative values for areas measuring 100×100 m^2 (Müntze, pers. comm. 1987). The pronounced variations of h_t and n_c and the type of tree are evident from Figs. 4.16 and 4.17.

The tree height varied between 6 and 25 m. Areas covered by short young trees ($h_t < 1.5$ m) were considered in the calculations only by an increased roughness length. The frequency distribution of h_t over the area investigated is shown in Fig. 4.18. The mode for most trees is around 20 m, whereas the mean

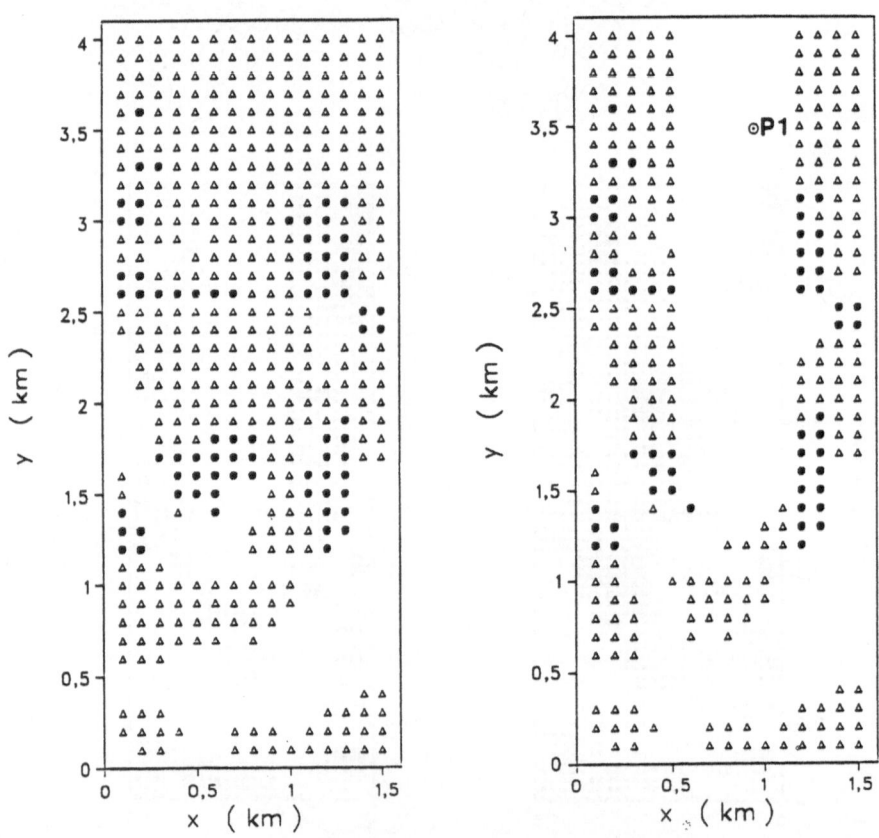

Fig. 4.17. Areal distribution of tree type prior to deforestation (*left*) and thereafter (*right*). *Circles* Deciduous forest; *triangle* coniferous forest. P1 indicates the location of a selected site

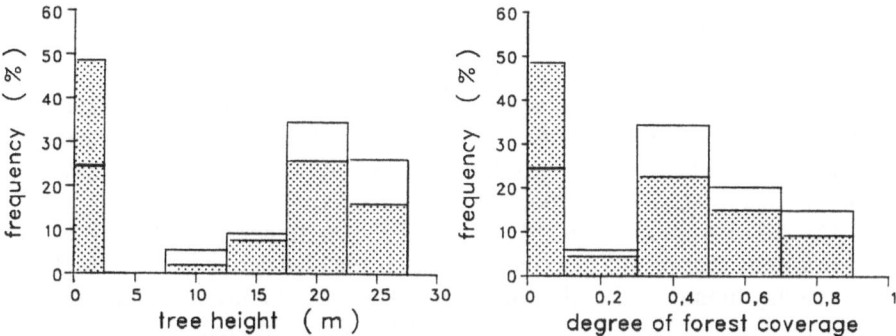

Fig. 4.18. Frequency distribution of tree height h_t (*left*) and stand density n_c (*right*). The *dotted area* indicates the situation after deforestation

height was calculated as 15.7 m. The stand density has a mode of $n_c = 0.4$ (Fig. 4.18) with a maximum of $n_c = 0.8$. This results in an average stand density of 39%. During deforestation, more large trees were cut down and the mean stand density was reduced to 26%. The vertical profiles of leaf area density for conifers and deciduous trees, described in Section 2.1, were also used in the present simulation.

Land utilization was subdivided into four types: forest, young plantation, open ground and concrete runway. For each of these types characteristic values for the various input parameters were defined (Table 4.4).

Because for the concrete runway the supply of moisture from the soil is not possible evaporation is excluded by setting $Q_V = 0$.

The DWD study concluded that the most pronounced influences on the local climate would be encountered during situations with large-scale low wind velocity. Consequently, a weak large-scale wind with $u_g = 2.5 \text{ m s}^{-1}$ and $v_g = 1.5 \text{ m s}^{-1}$ is assumed. In this study, the x-axis is oriented in the west-east direction, while the y-axis is oriented south-north. The thermal stratification of

Table 4.4. Site parameters for different land use

		Open ground	Forest	Young plantation	Runway
v_s	$(\text{m}^2 \text{ s}^{-1})$	4×10^{-7}	7×10^{-7}	5×10^{-7}	1×10^{-6}
λ_s	$(\text{W m}^{-1} \text{K}^{-1})$	1.20	2.00	1.40	3.00
$T(-0.5 \text{ m})$	(K)	290	290	290	290
W_{FC}	(m)	0.005	0.02	0.01	
α^*	$(\text{kg m}^{-3} \text{s}^{-1})$	3×10^{-3}	6×10^{-3}	4×10^{-3}	
a_g/a_t		0.25	0.15	0.20	0.10
ε		0.98	0.98	0.98	0.98
z_0	(m)	0.04	0.20	0.30	0.01

the simulation at the initial time corresponds to that of the standard atmosphere and a relative humidity of 30% is set. The radiation conditions assumed are those of a cloudless autumn day (September 23rd).

The grid intervals in the horizontal direction are $\Delta x = \Delta y = 100$ m. As in the calculations described so far, the vertical grid is non-equidistant. The locations of the grid levels above ground are 0, 2, 4, 6, 10, 15, 20, 25, 40, 70, 100, 150, 200, 250, 300, 400, 500, 700, 1000, 1500, 2000 and 2500 m.

The calculations start at 2100 LST and are continued to 2400 LST of the following day. The results described here cover the 24-h period from midnight of the first day to midnight of the second day. The situation prior to deforestation will be referred to hereafter as case 1, and the situation after deforestation as case 2. The selected site P1 has the coordinates $x = 0.9$ km and $y = 3.5$ km.

The direct oncoming solar radiation is attenuated while penetrating the stand and reaches the ground only in a greatly reduced form. At position P1 at 1200 LST only 130 W m^{-2} are available to heat the ground. After deforestation, Q_S attains a maximum value of 620 W m^{-2} (Fig. 4.19). Because of the low turbulence between 0 and 2 m, the turbulent sensible and latent heat fluxes are of little importance compared with the soil heat flux. The amplitude of Q_S, and thus also of the other energy fluxes, is controlled to a large degree by n_c, $b(z)$ and h_t. As these three parameters vary considerably throughout the area investigated at virtually every grid point, a different maximum value is simulated; however, this lies between the two extremes discussed here.

The larger diurnal variation of Q_S also leads to a strong fluctuation in the surface temperature with time (Fig. 4.20). Compared with case 1, the largest temperature increase in case 2 is 12.8°C, whereas the minimum difference simulated is -3.4°C. The slower rate of cooling in case 2 is caused by a plume

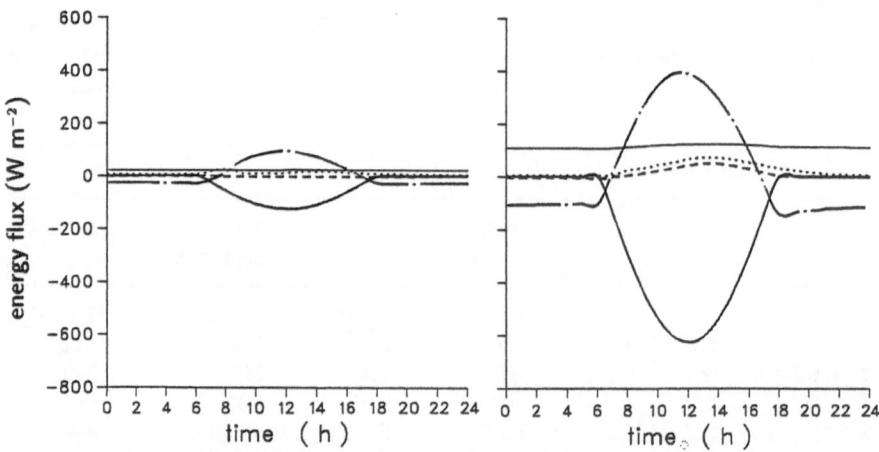

Fig. 4.19. Energy fluxes at ground level at $x = 0.9$ km and $y = 3.5$ km prior to deforestation (*left*) and thereafter (*right*). *Heavy solid line* Q_S; *solid line* Q_L; *dashed and dotted line* Q_B; *dashed line* Q_H; *dotted line* Q_V

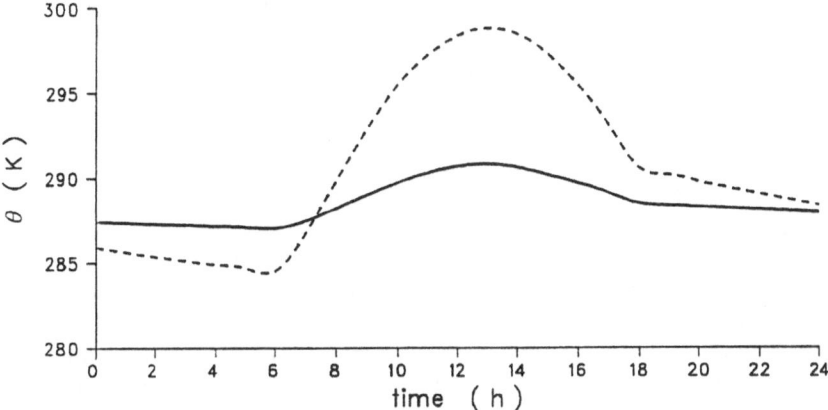

Fig. 4.20. Diurnal surface temperature variation at $x = 0.9$ km and $y = 3.5$ km prior to deforestation (*solid line*) and therefore (*dashed line*)

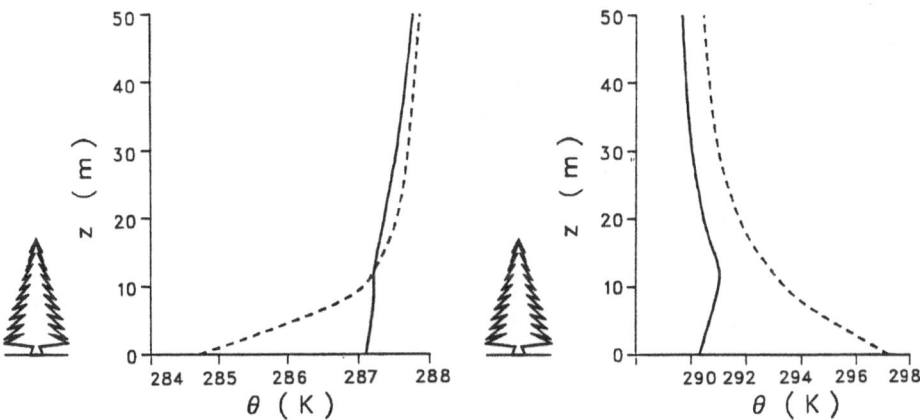

Fig. 4.21. Vertical profiles of potential temperature at $x = 0.9$ km and $y = 3.5$ km at 0600 LST (*left*) and 1200 LST (*right*) prior to deforestation (*solid line*) and thereafter (*dashed line*)

of warm air emanating from the concrete runway, which is heated markedly during the daytime.

Characteristic vertical profiles of the potential temperature in the two cases discussed show an almost constant value of θ in the forest at night with a weak cooling in the upper canopy (Fig. 4.21). At the same time, a pronounced surface inversion is developed when the sheltering forest cover has been removed. This is rapidly diminished during the early morning hours and at midday a strong stratification is simulated for the same site.

With trees present, the main heating occurs in the upper third of the canopy, provided that the crowns are sufficiently dense. Above this, a slightly unstable stratification is simulated, while in the trunk zone the potential temperature increases with height.

The vertical cross section of θ at around midday (Fig. 4.22) shows the pronounced heating of the runway. The prevalent wind transports the warmed air eastward and the warm air plume is still notable at $x = 1.6$ km.

In the horizontal section for the surface temperature the great differences between forest and open ground are clearly evident (Fig. 4.23). Even in the transition zones between stands of different height and density, large temperature gradients are developed. As expected, the air close to the ground is warmest around midday over open ground. A similar pattern is observed after deforestation with the sole exception that the highest temperatures are now encountered over the runway itself.

Figure 4.24 illustrates the difference between the calculated ground temperatures at 1200 LST $[\theta_0 \text{ (case 2)} - \theta_0 \text{(case 1)}]$. The differences are restricted to the deforested areas and the largest deviations are developed over the concrete

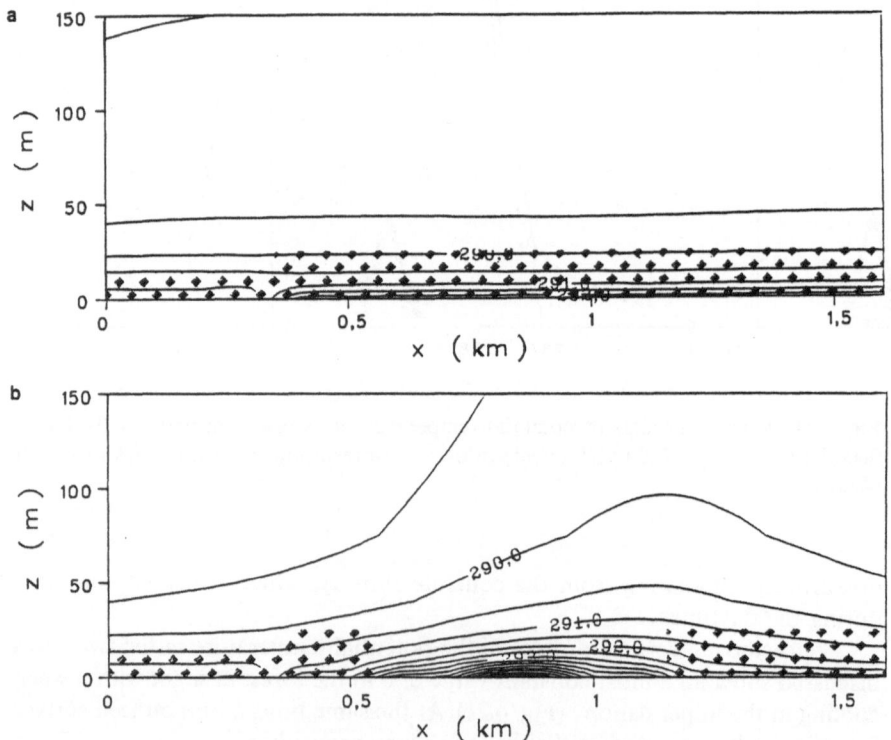

Fig. 4.22. Vertical section of potential temperature at 1200 LST at $y = 3.5$ *km*. **a** Prior to deforestation and **b** thereafter (K, intervals 0.5 K). The *dotted area* indicates the canopy

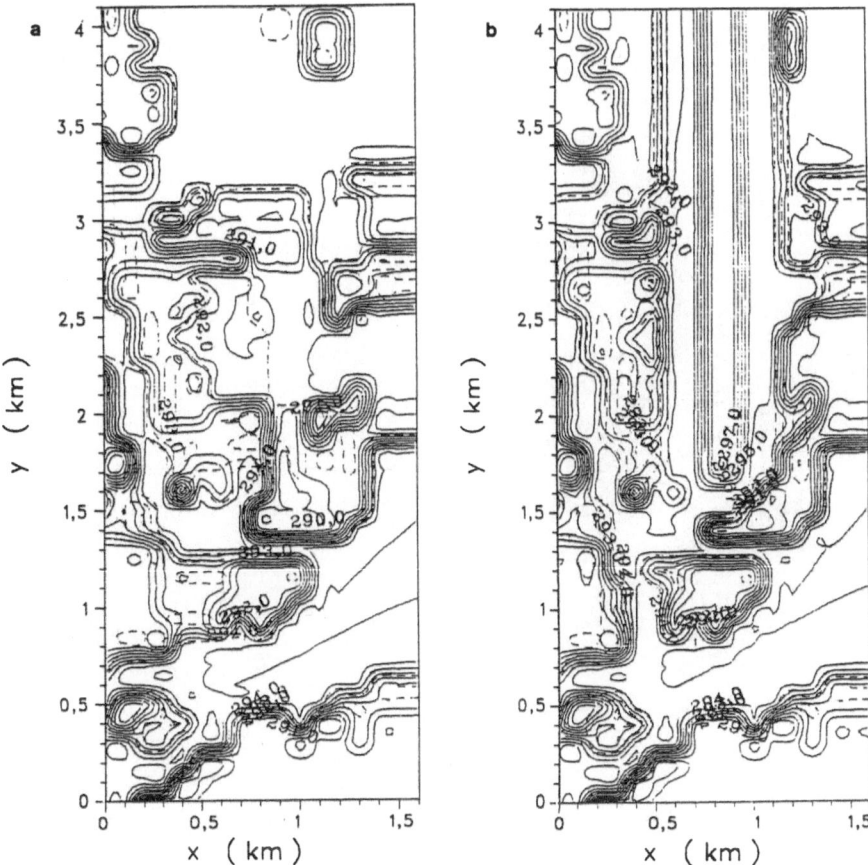

Fig. 4.23. Horizontal sections of surface temperature at 1200 LST. **a** Prior to deforestation and **b** thereafter (K, intervals 0.5 K)

runway. In the clearing into which the overhead powerlines were relocated, 2.5°C higher temperature was simulated at this time.

The diurnal variation of the relative humidity in case 1 shows a rather low amplitude (Fig. 4.25). As a result of the strong insolation over open ground in case 2, the relative humidity at midday is strongly reduced, but increases again during the evening hours. Vertical sections of the differences in relative humidity (Fig. 4.26) show a relatively drier atmosphere during the daytime, corresponding to the calculated temperature pattern. The stronger nocturnal cooling of the grass-covered safety strip along the runway leads to an increase in relative humidity by up to 25%. This notably increases the probability of ground fog formation in this zone, as pointed out in the DWD study.

The increased roughness close to the ground, resulting from the existing forest, notably reduces the wind speed (Fig. 4.27). Whereas in case 1 values

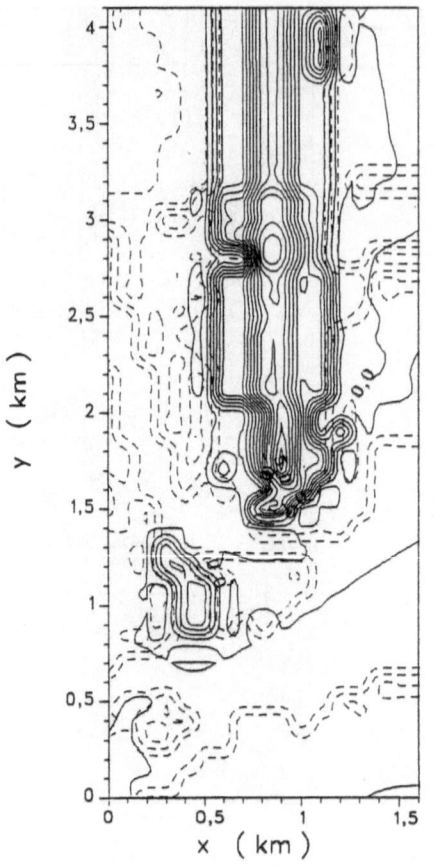

Fig. 4.24. Horizontal section of surface temperature difference at 1200 LST (K, intervals 0.5 K)

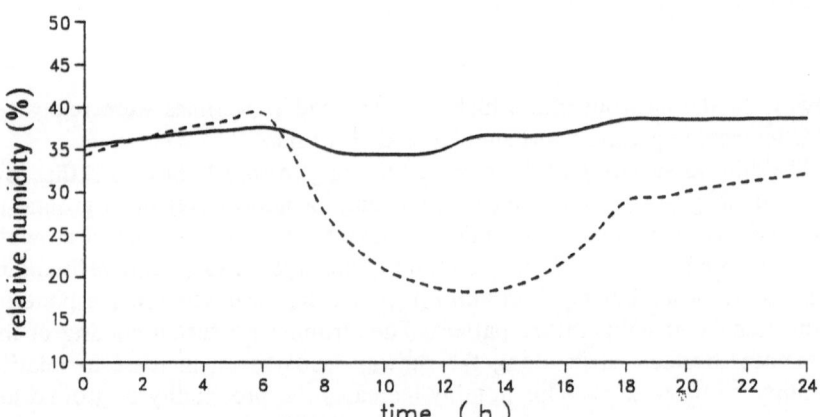

Fig. 4.25. Diurnal variation of surface relative humidity at $x = 0.9$ km and $y = 3.5$ km prior to deforestation (*solid line*) and thereafter (*dashed line*)

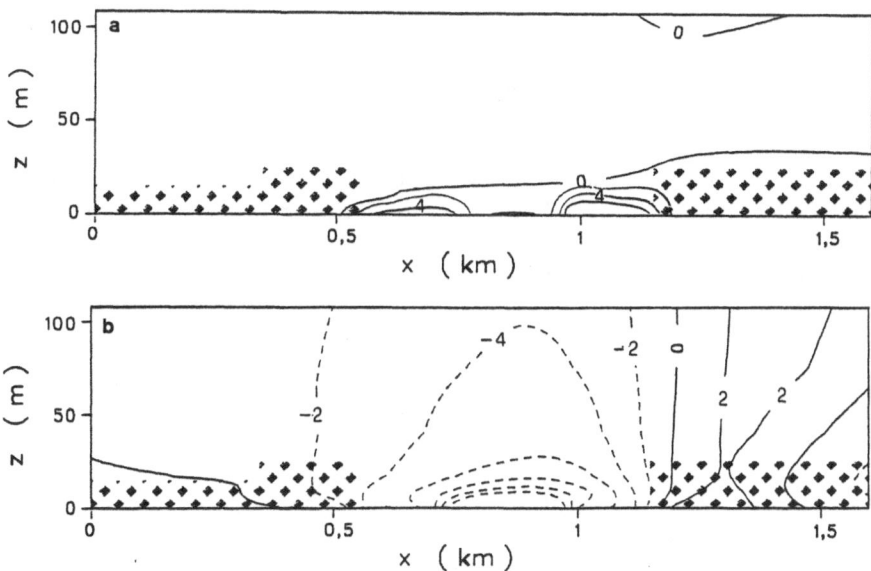

Fig. 4.26. Vertical sections of relative humidity differences (prior to and after deforestation). **a** Night and **b** daytime (%, intervals 2%). The *dotted area* indicates the canopy

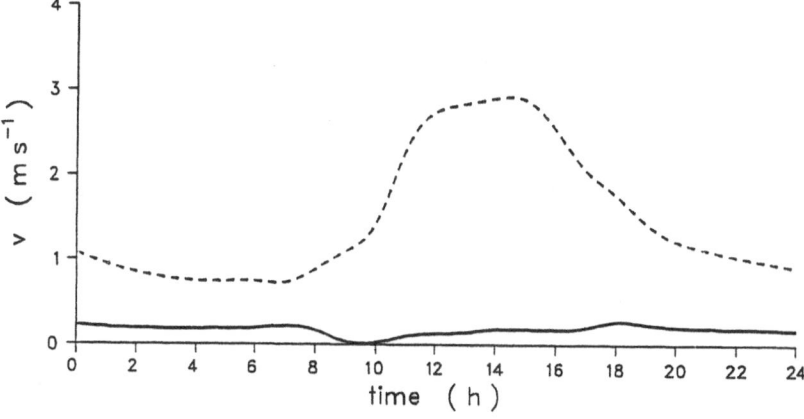

Fig. 4.27. Diurnal variation of wind speed 6 m above ground at $x = 0.9$ km and $y = 3.5$ km prior to deforestation (*solid line*) and thereafter (*dashed line*)

above 0.3 m s^{-1} are rarely simulated at point P1, V shows a pronounced diurnal variation along the western runway after deforestation with the maximum values of up to 3 m s^{-1}. These high velocities are encountered during those times when, under unstable stratification, strong vertical turbulent mixing links the low level air flow closely to the geostrophic wind. With increasing thermal stratification of the atmosphere close to the ground, the wind speed tends to diminish. As the conditions of stratification in the forest are exactly opposed to those over open ground with a more stable stratification during daytime, a corresponding behavior of the wind speed is simulated with lower values during the day.

In Fig. 4.28 the diurnal wind speed variation 6 m above ground at $y = 3.5$ km is presented. It clearly shows the effect of the clearing with large values of V. Along the eastern boundary of the clearing, at $x = 1.2$ km, the air flow is slowed abruptly and thus it not able to penetrate deeper into the stand. On the opposite side there is a small acceleration out of the forest.

In a vertical section of V at midday (Fig. 4.29), the acceleration of the wind as it moves from forest to open ground is clearly evident, as is the retardation on

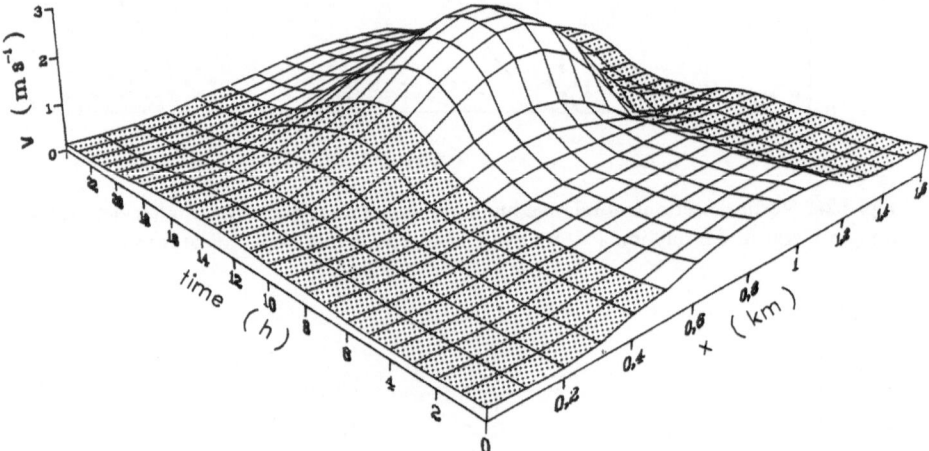

Fig. 4.28. Diurnal variation of wind speed 6 m above ground at $y = 3.5$ km. The *dotted area* indicates the location of the stand

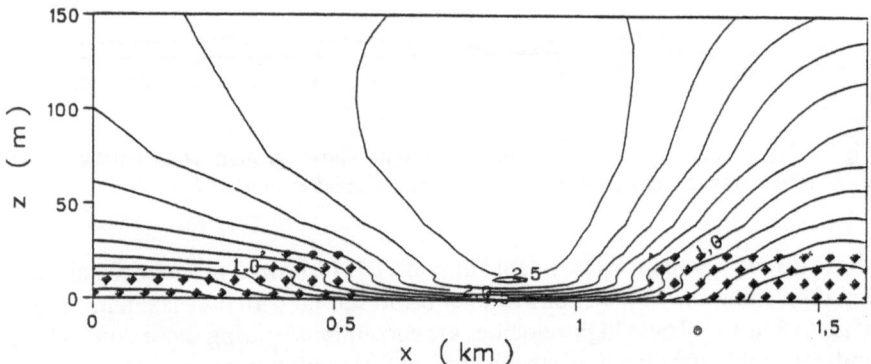

Fig. 4.29. Vertical section of wind speed at 1200 LST at $y = 3.5$ km (m s^{-1}, intervals 0.25 m s^{-1}). The *dotted area* indicates the location of the stand

entering the stand on the opposite side. This shows that at this position the influence of the clearing extends for 200–300 m into the forest. When the wind penetrates the stand, a secondary velocity maximum is established at $x = 1.2$ km. This is also clearly developed in the vertical profile of V at 0600 LST (Fig. 4.30). It disappeas later (1200 LST), only to reappear during the evening hours.

Horizontal cross sections of wind at 2 m above ground at 0600 LST show the pronounced channeling effect of the existing clearing (Fig. 4.31a–d). While over the open ground to the south of the runway southwesterly winds prevail, these change from southwesterly to southerly in the clearing itself. During the daytime the wind close to the ground increases again (Fig. 4.31c–d), at the same time turning more to a westerly direction. The pronounced warming of the open ground leads to local circulation systems which supply cooler air from the stand.

This phenomenon of a 'forest wind' is clearly evident from the simulated patterns of either side of the clearing (Fig. 4.32). During the night a horizontal pressure gradient between the clearing and the forest is established as the result of the radiative cooling over the field, and this leads to flow into the stand. During the day the situation is reversed and the air now flows out of the forest. This rather weak circulation system is superimposed on the large-scale wind and, as a result, may only be deduced from the retardations and accelerations of V along the stand margins.

Observations of such local thermal wind systems during the daytime have been reported by Berg (1947) and Geiger (1961). Because of the lower temperature differences between forest and open ground during the night, a nighttime counterpart of the forest wind is not observed (Geiger 1961). From a maximum simulated wind speed of $0.3\ \mathrm{m\,s^{-1}}$ for the forest wind and $0.1\ \mathrm{m\,s^{-1}}$ for the opposite circulation, this is quite obvious.

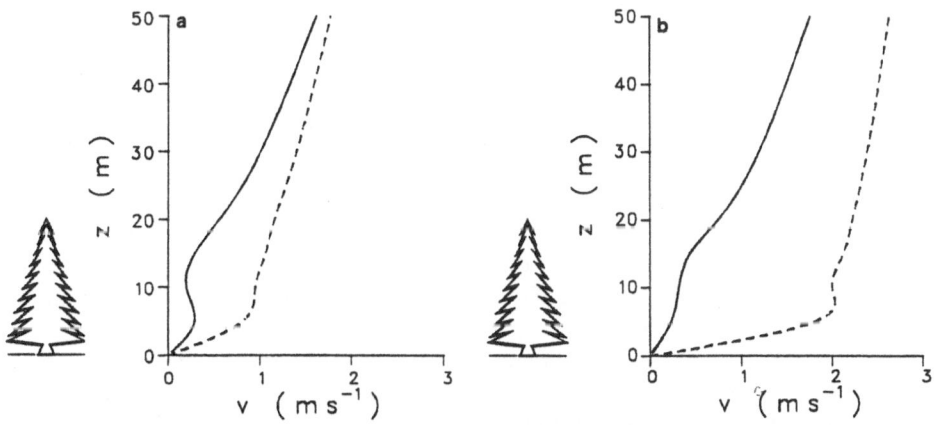

Fig. 4.30. Vertical wind speed profiles at $x = 0.9$ km and $y = 3.5$ km. **a** At 0600 LST; **b** at 1200 LST prior to deforestation (*solid line*) and thereafter (*dashed line*)

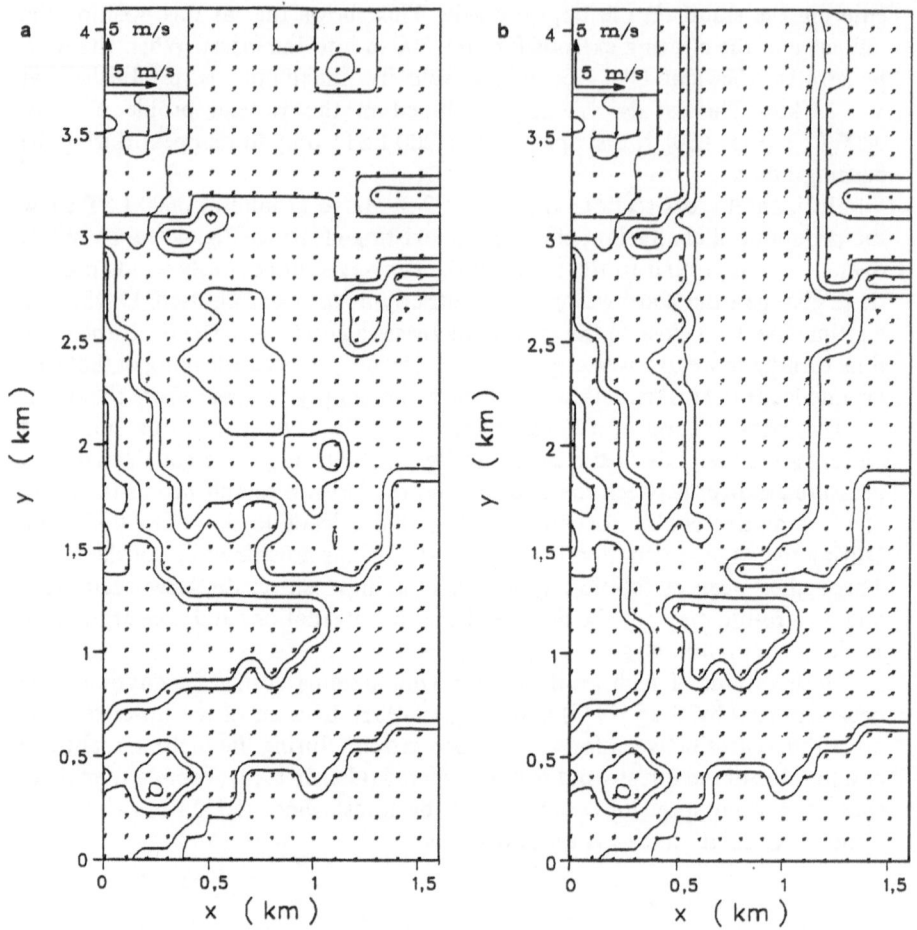

Fig. 4.31. Horizontal sections of the air flow 2 m above ground. **a** At 0600 LST for case I; **b** at 0600 LST for case II; **c** at 1200 LST for case I; **d** at 1200 LST for case II

When the distribution of differences of V for both simulations during the day and night is considered (Fig. 4.33), it is found that differences are virtually retricted to only the deforested areas. Only east of the runway does the air flow penetrate deeper into the stand during the day, leading to larger differences. At the level selected, the largest differences are 0.5 m s^{-1} during the night and up to 2.4 m s^{-1} during the day.

Although the DWD has already measured wind speeds over the western runway, so far they have been evaluated in a rather exemplary manner only (Beffert, pers. comm. 1987). On October 12th 1986, the Frankfurt area was within a large-scale southwesterly flow, i.e. under the same weather conditions as those assumed in the simulations. At an observation station, located in the

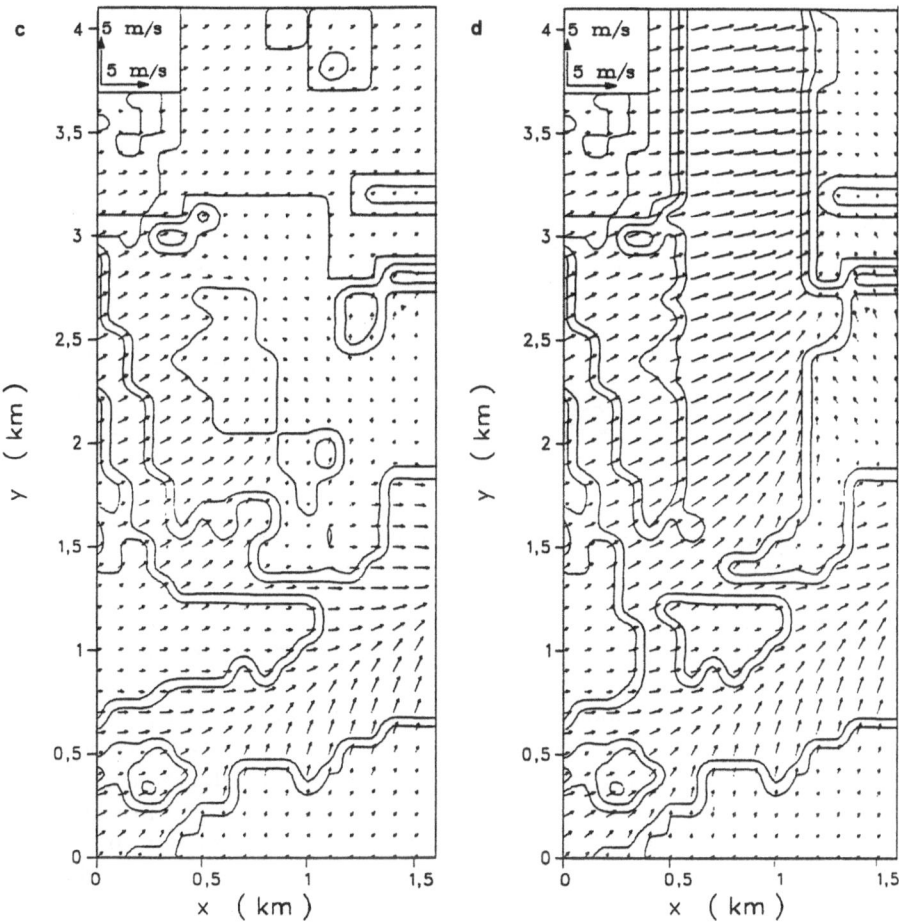

Fig. 4.31. c and d

northern part of the western runway, wind speeds between 3 and 5 m s^{-1} were registered. These values are about double the simulated ones. Despite this quantitative difference, which can be ascribed to the selected geostrophic wind speed, the observed diurnal variation of the wind direction is rather similar to the simulated one (Fig. 4.34). At night, the air flow becomes channeled by the clearing, whereas after sunrise it turns more into the direction of the large-scale wind. The observations show that at 1400 LST the wind can even blow from the northwest, whereas the simulations only resulted in a westerly wind. This channeling effect was also noted by Geiger (1950) in his field measurements on the influence of forest clearings. He also found that for an oblique impacting

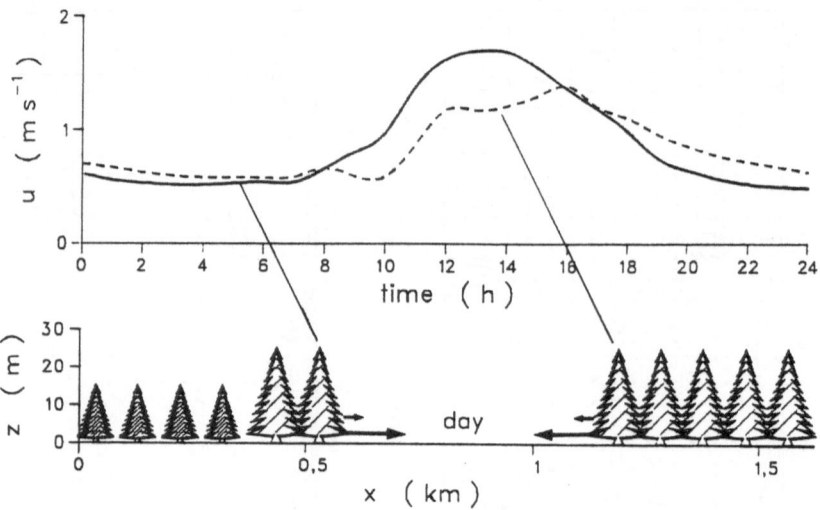

Fig. 4.32. Diurnal variation of $\bar{u}$ component at 4 m above ground at $y = 3.5$ km (*solid line*, western edge of clearing; *dashed line* eastern edge)

large-scale wind, the velocity maximum was shifted to the downwind side of the cut, as was also shown by the simulations (Fig. 4.31b).

The vertical cross section of the temperature distribution clearly showed the warm air plume, caused by the overheated runway, being transported eastward by the larger-scale wind. The corresponding cross section of the vertical velocity (Fig. 4.35) illustrates a zone with upward winds above 0.2 m s^{-1} in which the warm plume is advected above the treetops to the east. The 20-m-deep clearing acts like a small valley leading to the descent of flow on the one side and to its ascent on the opposite side. The increasing instability of the boundary layer close to the ground during the subsequent 2 h resulted in an increase of the vertical velocity to above 1 m s^{-1}. During the night, however, a vertical velocity of 0.05 m s^{-1} was not exceeded. The relatively large wind speed during the daytime at treetop height in combination with the unstable stratification encountered at the same position led to a high rate of transformation of the kinetic energy of the mean flow and of its internal energy into turbulent kinetic energy (Fig. 4.36). During both the day and night the maximum is developed within the upper third of the stand. Within the underlying zone a rapid decrease of E is found as a result of the stable stratification and the pronounced retardation of V. In contrast to this, the turbulent kinetic energy is rather low and constant with height over the deforested area. Even during the day, E does not reach the maximum vaues simulated in case 1. In the vertical section (Fig. 4.37) the treetops and the stand margins can be identified as preferred production zones of turbulent kinetic energy.

The effects of deforestation along the western runway on the local and regional climate may be estimated by comparing the simulation results for both

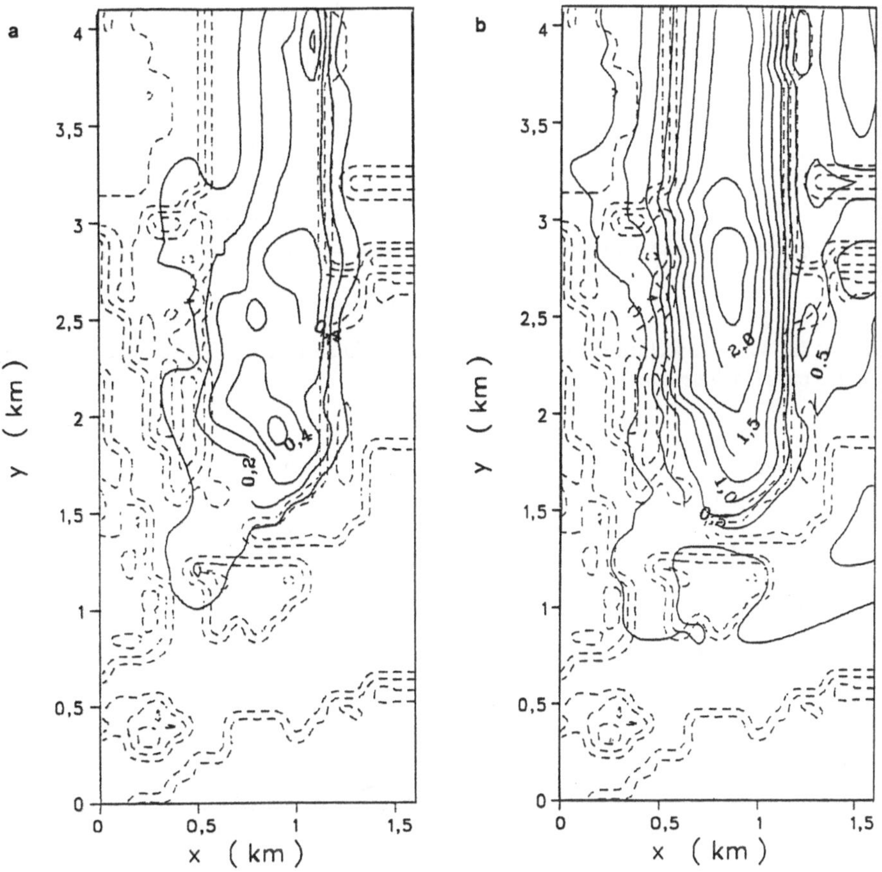

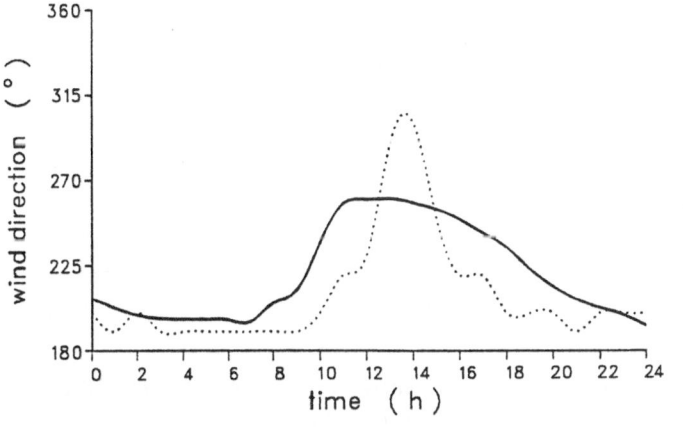

Fig. 4.33. Differences in wind speed 2 m above ground. **a** At 0600 LST (interval 0.1 m s^{-1}) and **b** 1200 LST (interval 0.25 m s^{-1})

Fig. 4.34. Diurnal variation of wind direction 2 m above ground at x = 1.0 km and y = 3.9 km (*solid line* simulated; *dotted line* observed)

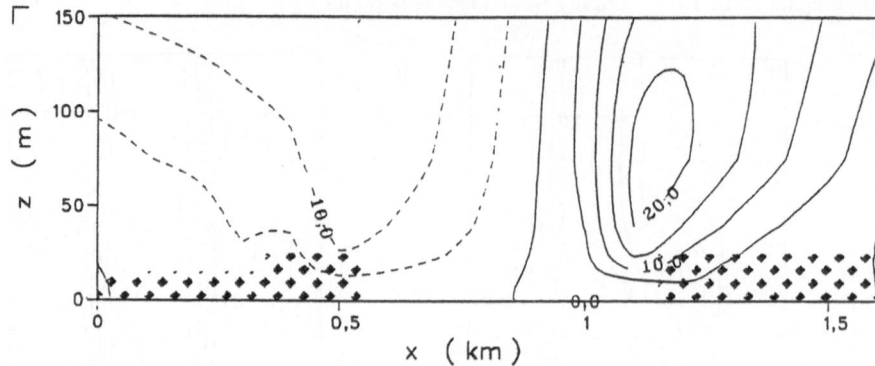

Fig. 4.35. Vertical section of vertical velocity at 1200 LST at $y = 3.5$ km (cm s^{-1}, intervals 5 cm s^{-1}). The *dotted area* indicates the canopy

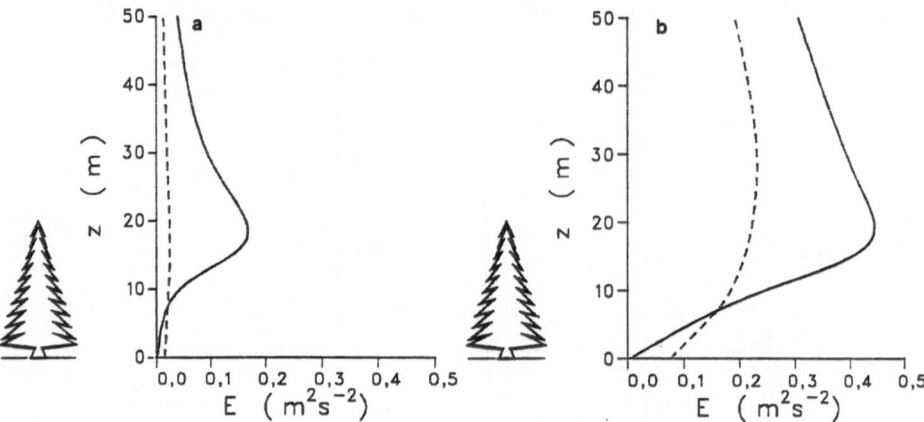

Fig. 4.36. Vertical profiles of E at $x = 0.9$ km and $y = 3.5$ km. **a** At 0600 LST and **b** 1200 LST (*solid line* prior to deforestation; *dashed line* thereafter)

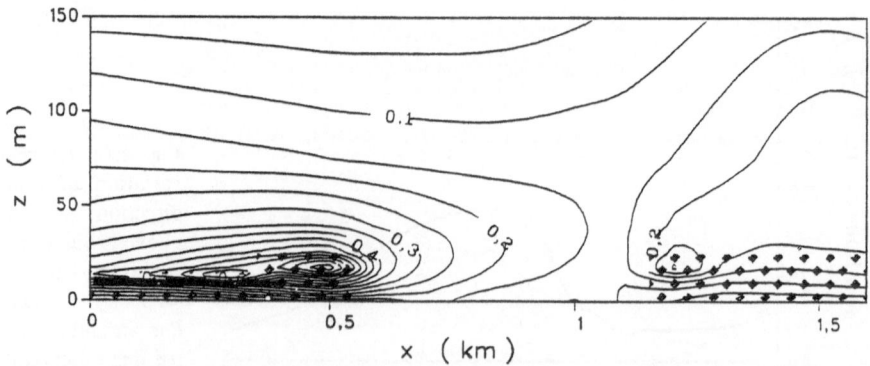

Fig. 4.37. Vertical section of turbulent kinetic energy at $y = 3.5$ km at 1200 LST (m^2 s^{-2}, intervals 0.05 m^2 s^{-2}). The *dotted area* indicates the location of the stand

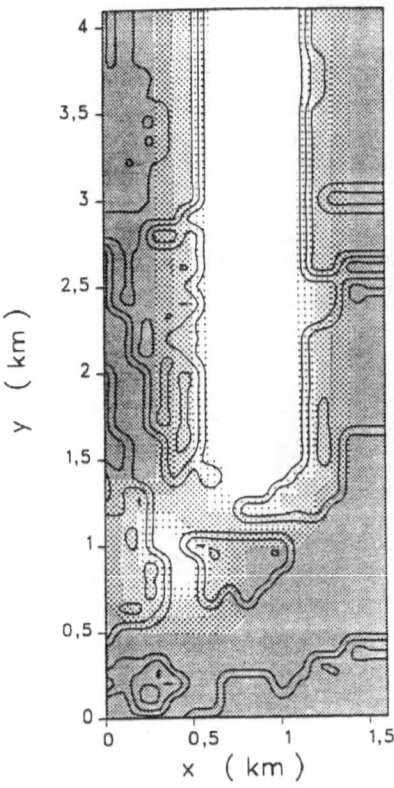

Fig. 4.38. Arrangement of areas needed for calculation of climatic changes (distance from the deforested areas, *unshaded areas* indicate the clearing)

cases. For this purpose, the mean variations of temperature, humidity and wind speed over the 24-h simulation period at 2 m above ground were calculated for the shaded areas shown in Fig. 4.38. Grid points over which the forest had been cleared were not taken into consideration.

The following mean differences were simulated for the variables in question: temperature, $-0.04°C$; relative humidity, -0.01% and, wind speed, $0.05 \, \text{m s}^{-1}$.

Such small differences may be calculated but cannot be measured in the field. This leads to the conclusion that regional climate around the Frankfurt airport is not affected by the deforestation.

The values given above are controlled to a large extent by the arbitrarily chosen size of the area of averaging, in this case the simulation area.

In order to determine the influence of the distance from the clearing on the climatic parameters considered here, another evaluation of the simulation results was carried out. The variations were determined for strips of 100-m width parallel to the boundaries of the deforested zones. The results for temperature and wind speed are shown in Fig. 4.39. The mean warming over the deforested area was found to be 1°C, while the increase in the mean wind velocity is about $0.75 \, \text{m s}^{-1}$. With increasing distance the changes of V became smaller. The mean

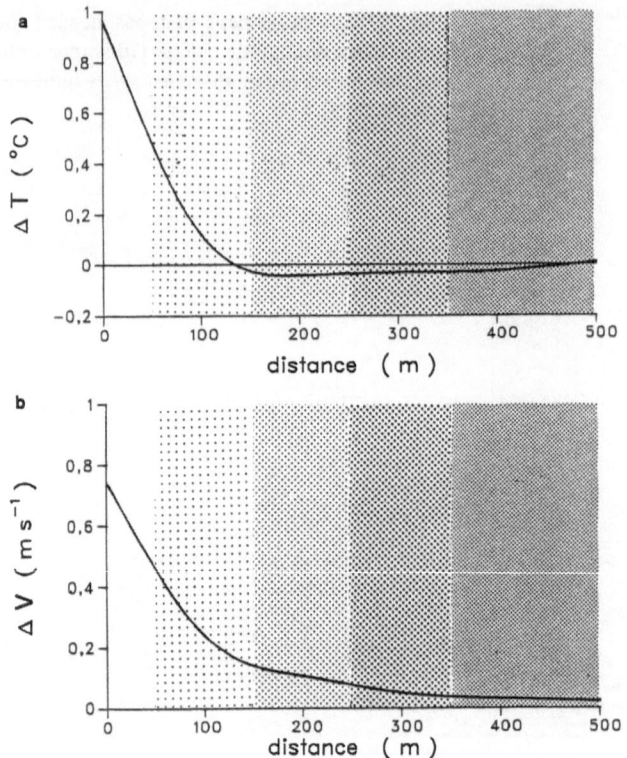

Fig. 4.39. Mean change vs. distance from deforested areas. **a** Temperature; **b** wind speed. The intensity of shading relates to that shown in Fig. 4.38

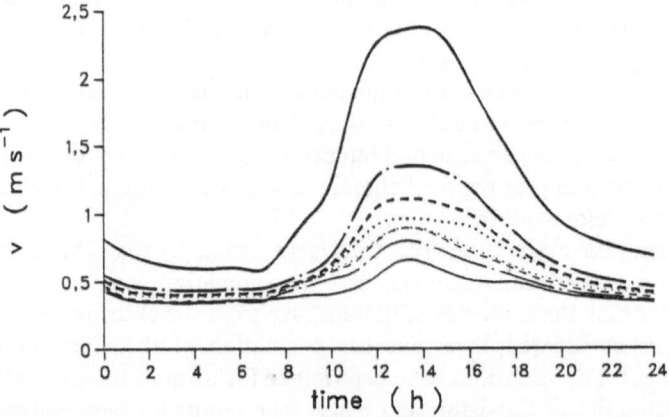

Fig. 4.40. Mean diurnal variation of V at 4 m above ground vs. distance from deforested areas. *Heavy lines* prior to deforestation and *thin lines* thereafter; *solid line* 0 m; *dashed and dotted line* 100 m; *dashed line* 200 m; *dotted line* 300 m)

temperature difference attains another extreme value of $-0.1°C$ at 200 m from the edge of the forest, and decreases thereafter.

The differences in T and V between the two simulations already suggested that the mean differences in the variables outside the cleared area would be rather small. A comparison of the mean diurnal variations in wind speed over areas of varying distance from the cutting with the same areas prior to deforestation again clearly illustrates how quickly the influence of the change in ground cover diminishes (Fig. 4.40).

4.3.2 Effects of Complete Deforestation of the Finkenbach Valley (Odenwald) on the Local Climate

The orography of an area can influence the air flow in many ways. Geiger (1961) distinguished between active and passive influences. The latter term implies the modification of the large-scale wind pattern by hills and valleys; among the effects observed are flow over and around isolated obstacles (Walmsley et al. 1982; Wooldridge et al. 1982), the formation of lee waves (Klemp and Lilly 1975) and the flow pattern in the lee of obstacles (Zimmermann 1969).

In wide elongated valleys, such as the Upper Rhine valley of southwestern Germany, the bordering mountain ranges lead to a channeling of the air flow to such an extent that the surface winds virtually all blow parallel to the axis of the valley (Wippermann and Gross 1981). For certain large-scale winds the influence of the orography may even lead to surface winds where directions are opposed to the geostrophic wind (Gross and Wippermann 1987). An explanation for such 'countercurrents' was presented by Wippermann (1984).

Active influences exerted by the structure of the area on the wind pattern are the formation of local wind systems as the result of horizontal temperature and pressure gradients. The variation in nocturnal cooling and daytime warming of the atmosphere close to the ground leads to the establishment of diurnally changing circulation systems. These include nocturnal cold air drainage and upslope winds around midday together with mountain and valley winds (Defant 1951).

Vegetation, e.g., taller trees, will modify these wind systems. The nature of the modifications is rather difficult to determine by field measurements. Bergen (1969) described cold air drainage over a forested slope, but was unable to draw any conclusions to the quantitative and qualitative nature of such changes, because he did not have any measurements for an unforested case at his site. Only a numerical model, as described here, will give some deeper insight into the effects of such changes.

The area for this numerical study is the Finkenbach valley of the southern Odenwald mountains, some 15 km northeast of Heidelberg (Gross 1987). The location and a perspective view of the selected landscape are shown in Fig. 4.41. The valley orographically controls an area of 7×17 km². It is of a highly diverse

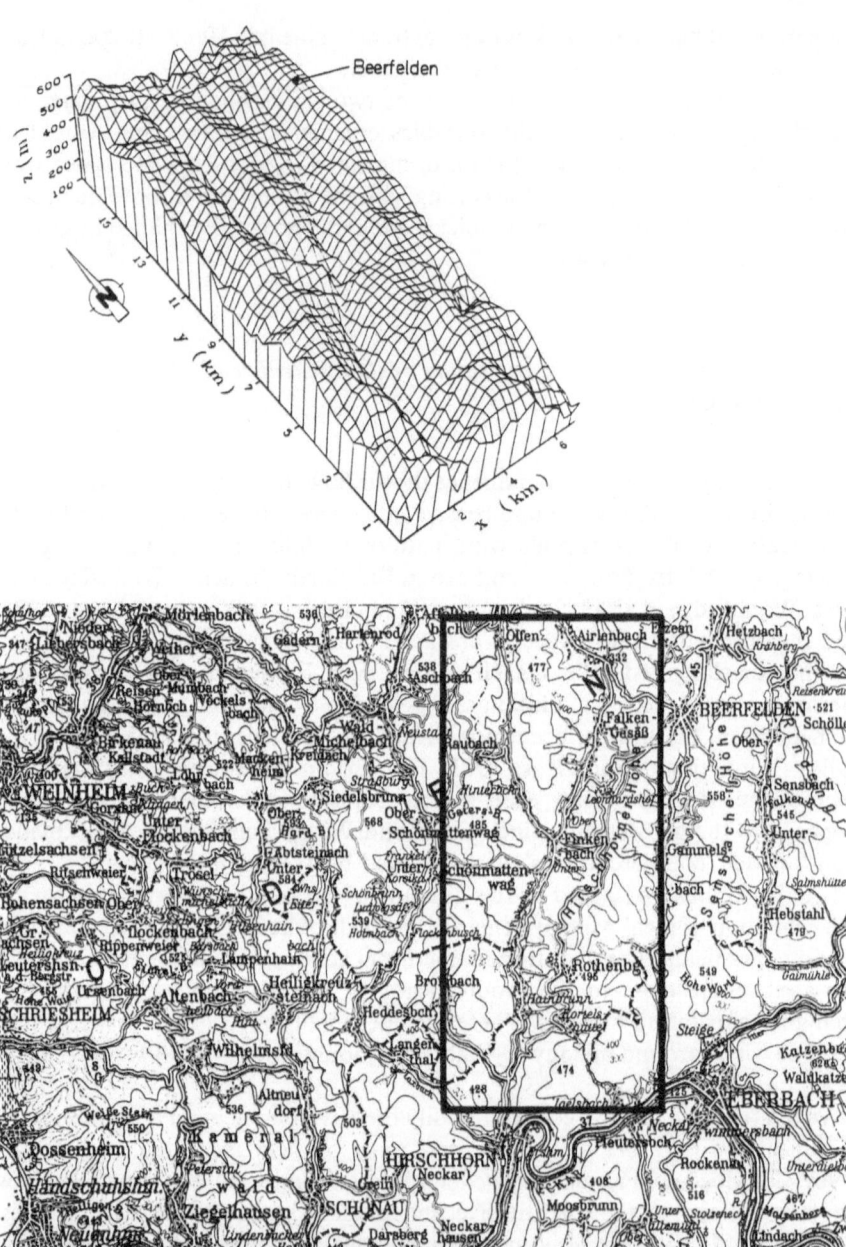

Fig. 4.41. Map and perspective view of the Finkenbach valley

orography with slopes typically of 10°–15° and with a maximum of 28°. At its southern end, the area is surrounded by 400- to 500-m-high hills.

The ground cover is rather heterogeneous, ranging from dense forest to agricultural fields. Trees covering the valley slopes account for about 80% of the area while only a 100- to 200-m-wide strip along the valley bottom is unforested. The stand consists of trees of different age and height with 50% spruce, 30% fir, 10% oak and 10% beech. According to information from the local forestry office, a typical average stand height of 20 m may be assumed with trees of optimum positions attaining heights of up to 35 m.

To achieve an impression of the forest density in the Finkenbach valley, the total area shown in Fig. 4.41 was subdivided on a grid with $300 \times 300 \, m^2$ intervals. For each of these units the portion covered by forest was determined from topographic maps. The results are shown in Fig. 4.42a, where the density of shading is proportional to n_c.

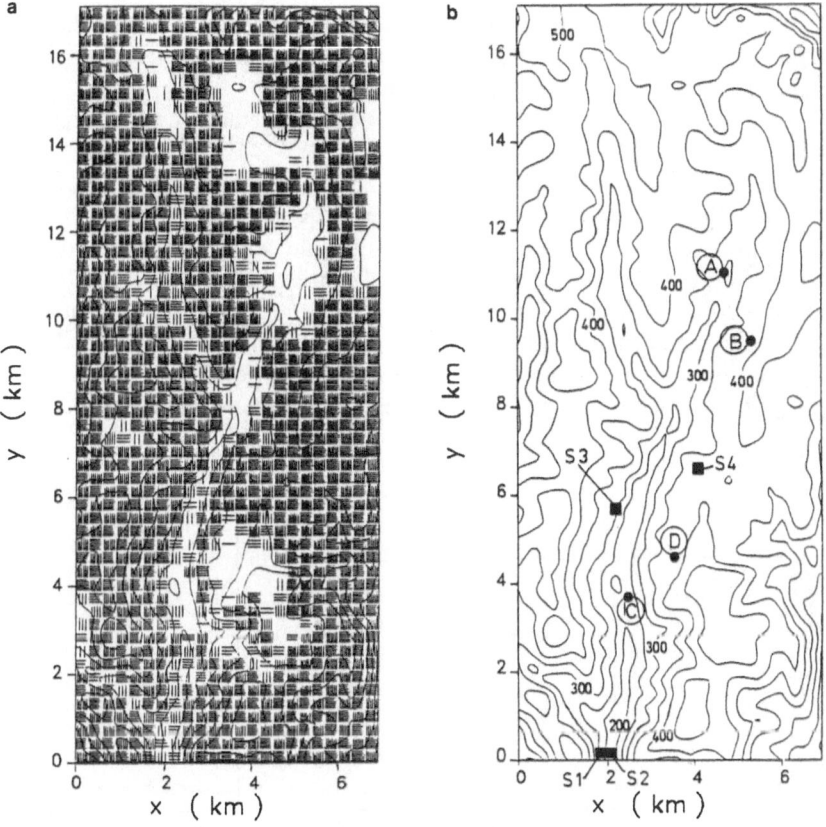

Fig. 4.42. Forest distribution in the Finkenbach valley (*left*). The density of shading is proportional to stand density. Positions of DWD monitoring stations 1954–1964 (*right*)

The Finkenbach valley was selected for this study also because the DWD had observed minimum and maximum temperatures at 27 stations throughout the Odenwald between 1954 and 1964 (Schnelle 1972). This special recording network was established for an investigation of the different local climates of the region. At many points valley and mountain stations were set up concurrently in order to find out the influence of the elevation on the temperature distribution.

Four stations of the network system are located within the area influenced by the Finkenbach valley. Minimum and maximum temperatures were recorded at 1.6 m above ground at the following stations (Fig. 4.42b): station A, Falken-gesäß (272 m above mean sea level); station B, Leonardshof (412 m above mean sea level); station C, Ober-Hainbrunn (178 m above mean sea level) and station D, Rothenburg (428 m above mean sea level).

The records show that during the winter months the mean difference in minimum temperatures between stations B and A was 1.4°C and between stations D and C was 1°C. In both cases, the valley floor is therefore cooler than the adjoining slopes. This nocturnal temperature pattern corresponds to the climatological 'thermal belt' (Koch 1961).

As pointed out above, about 80% of the trees in the Finkenbach valley are conifers with a mean height of 20 m. The leaf area density and leaf area index used in the present simulation are shown in Fig. 4.43.

The large-scale input parameters and the site parameters are summarized in Table 4.5.

The horizontal grid intervals in both horizontal directions are 300 m. In the vertical direction non-equidistant intervals were used (Table 4.6) in order to achieve an optimum resolution close to the ground with a minimum of grid levels. The total simulation domain was thus represented by $28 \times 62 \times 16$ grid points. An additional five levels below the earth's surface allowed the calculation of the temperature distribution in the soil.

In order to study the influence of the forest on the formation of nocturnal cold air drainage and minimum temperatures, two simulations were carried out.

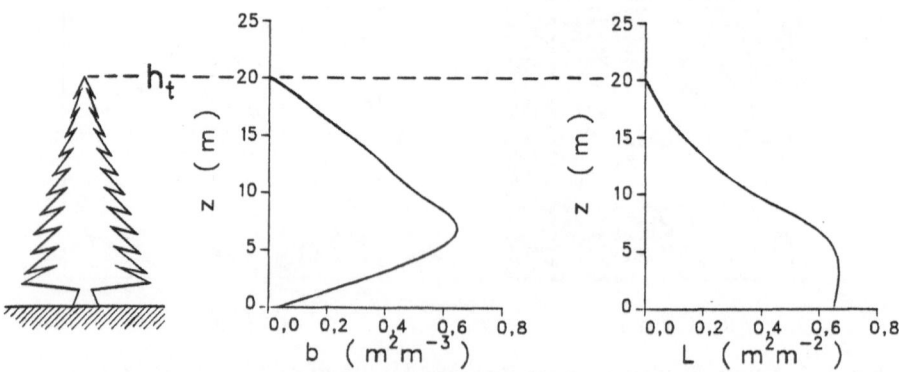

Fig. 4.43. Vertical profiles of leaf area density (*left*) and leaf area index (*right*)

Table 4.5. Meteorological input parameters and site parameters

u_g	0
v_g	0
φ	48° N
z_0	0.05 m
λ_s	1.25 $\mathrm{W\,m^{-1}\,K^{-1}}$
v_s	$4 \times 10^{-7}\,\mathrm{m^2\,s^{-1}}$
$T(-0.5\ \mathrm{m})$	277 K
a_g	0.25
ε_g	0.98
a_t	0.10
ε_t	0.98
k_c	0.60

Table 4.6. Location of grid levels in atmosphere and below ground (valid for $x = 2.1$ km, $y = 0$ km)

Grid level No. in atmosphere	Height (m)
16	1000
15	700
14	400
13	200
12	120
11	80
10	50
9	30
8	25
7	20
6	15
5	10
4	6
3	4
2	2
1	0
Below ground	
1	− 0.00
2	− 0.02
3	− 0.10
4	− 0.25
5	− 0.50

The present situation with dense stands in the Finkenbach valley is referred to as case I and the completely deforested situation is termed case II. In the latter it was assumed that the slopes are covered only by short grass. The simulations were started after sunset and the integration was carried out over a total of 6 h real time.

For better presentation of the results, all the variables were interpolated onto a common grid. This averaging led to an attenuation of the extremes.

During the assumed nocturnal situation, the ground is cooled because of the negative net radiation. The variation with time of the surface temperature for both cases at point S4 (cf. Fig. 4.42) is presented in Fig. 4.44. The grass-covered slope cools more rapidly than the ground sheltered by trees. After 3 h, a temperature difference of 1°C has been established with marginal changes thereafter. This value is controlled especially by the site parameters and the specific forest density. The temperature difference at the end of the simulation varied from place to place between 0 and 1.5°C.

The areal distribution of the surface temperature after 6 h is shown for both cases in Fig. 4.45. Isotherms are drawn only for the region belonging to the area of influence of the valley as shown by the dotted line. However, the calculation was carried out for the complete area.

The large variation in surface temperature for case I is caused by the heterogeneous structure of the stand. A minimum temperature of $-1°C$ was calculated here at $x = 2.4$ km and $y = 3.9$ km over the unforested valley bottom. This site is especially prone to cold air because: (1) outgoing longwave radiation over the unforested ground is large: (2) the rather sheltered position close to the taller trees reduces turbulent transport of warmer air from overlaying regions and (3) the unforested slope to the east supplies cooled air.

The temperatures simulated for the first grid level in the atmosphere at sites A–D are compared with the observations of the DWD. It should be pointed out, however, that point measurements at 1.6 m above ground are compared with averages for volumes measuring $300 \times 300 \times 1.8$ m^3. Despite this, the calculated temperature differences agree well with the observed values: between site A and B 1.1°C calculated vs. 1.4°C observed, and between C and D 0.9°C calculated vs. 1.0°C observed.

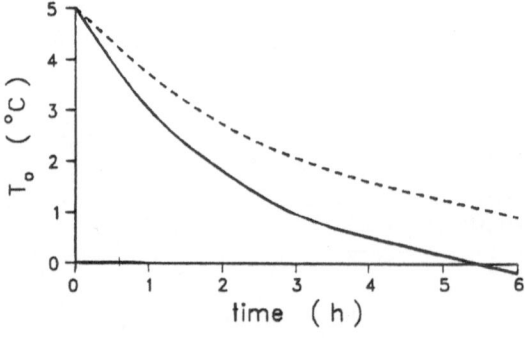

Fig. 4.44. Diurnal variation of surface temperature at site S4 (*dashed line* prior to deforestation; *solid line* thereafter)

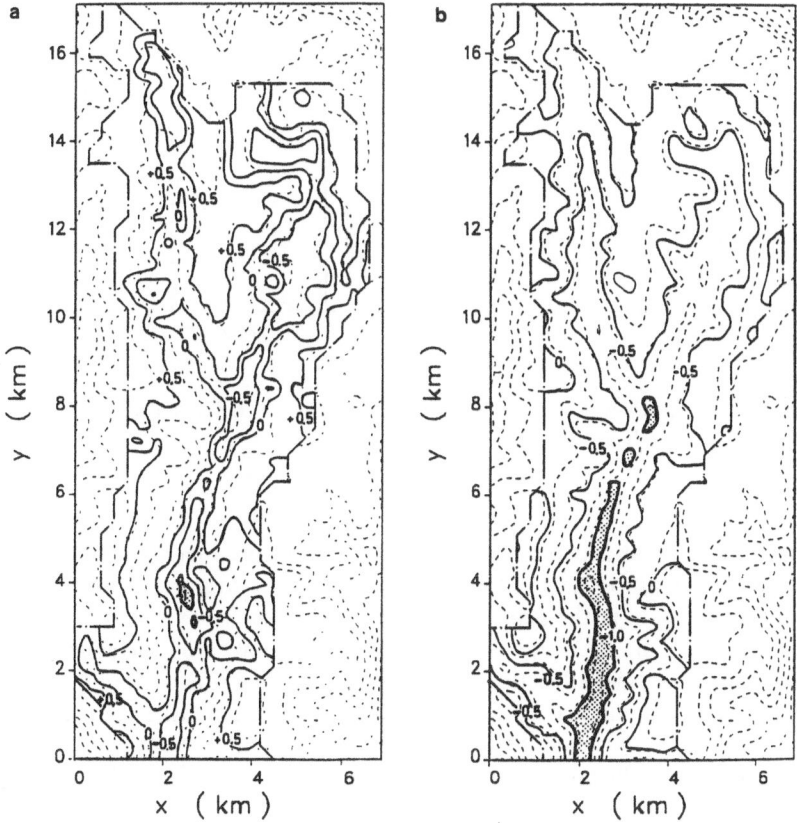

Fig. 4.45. Surface temperature **a** prior to deforestation and **b** thereafter ($T < -1°C$ is *shaded*). Terrain height is shown by *dashed lines*

After complete clearing of the forest a notably different temperature pattern is obtained (Fig. 4.45b). As a first approximation, the isotherms and contour lines of the orography are parallel with cold air in the lower parts of the area and relatively warmer slopes. The qualitative comparison of the two cases shows that the deforestation leads to notable reductions in surface temperature which, in turn, influence the newly established wind pattern.

With the longwave radiation scheme used here an atmospheric cooling rate of $0.14°C\,h^{-1}$ is calculated for the simulated temperature and humidity profiles at site S4 in case II. This value changes only slightly with height (Fig. 4.46a).

With trees present, a completely different pattern is developed. The divergence of the longwave radiation flux within the stand is negative with a pronounced maximum in the upper part of the canopy. The maximum cooling in this case is $22°C\,h^{-1}$ at a height of 15 m above ground.

The coldest region in the vertical temperature profile over unforested ground is close to the surface with an overlying inversion (Fig. 4.46b). In the forest the

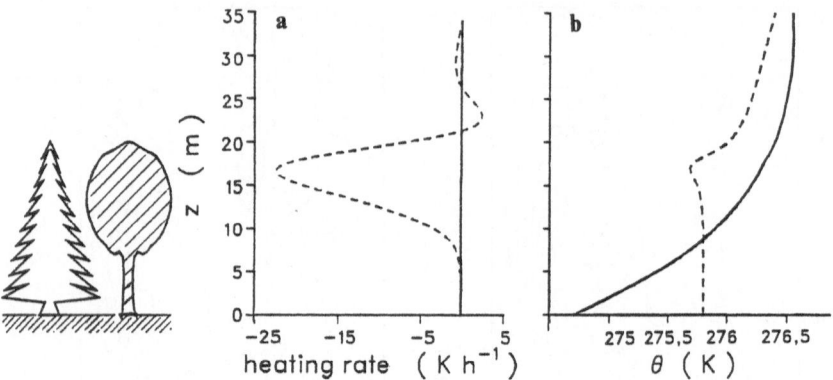

Fig. 4.46. Vertical profiles at site S4 (*dashed line* prior to deforestation; *solid line* thereafter). **a** Radiative heating rate; **b** potential temperature

minimum is located at a certain level with a temperature increase above and a neutral stratified atmosphere in the trunk zone below. However, the simulated temperature minimum in the stand is not as low as might be expected from the cooling rate of $22°C\,h^{-1}$. Compensating effects like an increased turbulent heat flux and stronger vertical advection will counteract such a drastic cooling. This may be shown by a numerical example. With an assumed potential temperature gradient of $0.1°C\,m^{-1}$ near the top of the canopy, a vertical velocity of $0.06\,m\,s^{-1}$ will be sufficient to compensate completely the effect of radiative cooling by the advection of warmer air.

Vertical cross sections of potential temperature for both cases at $y = 0$ km (Fig. 4.47) clearly illustrate the difference close to ground level. Within the stand, an almost neutral stratified atmosphere is observed with an inversion starting above the trees. After deforestation the temperature increase has already started at ground level. When the values of the isotherms are compared it is found that the valley atmosphere is warmer after deforestation than before. This may be explained by an increased vertical advection resulting from an increase in the horizontal convergence along the valley axis in case I. This point will be further discussed later.

A comparison of the temperatures for each grid point in the two cases is shown in Fig. 4.48. The solid line represents a perfect correlation. The surface temperature in the forest is about $1°C$ higher than over the deforested area. At grid points with lower stand densities (circles), the difference is much smaller. The same comparison carried out for a higher level, approximately coinciding with the crowns, leads to an entirely different pattern. The outgoing longwave radiation causes the temperature above the forest in case I to be about $1.8°C$ lower than in case II. It is interesting to note that the values for grid points with $n_c = 0$ are scattered around the dashed line and this indicates the complete correlation.

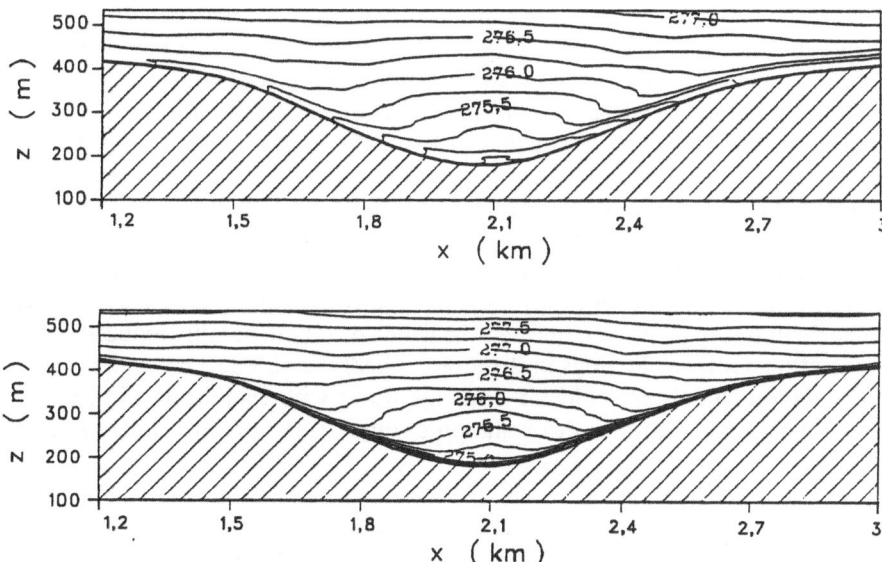

Fig. 4.47. Vertical sections of potential temperature at $y = 0$ km. *Top* Prior to deforestation and *bottom* thereafter

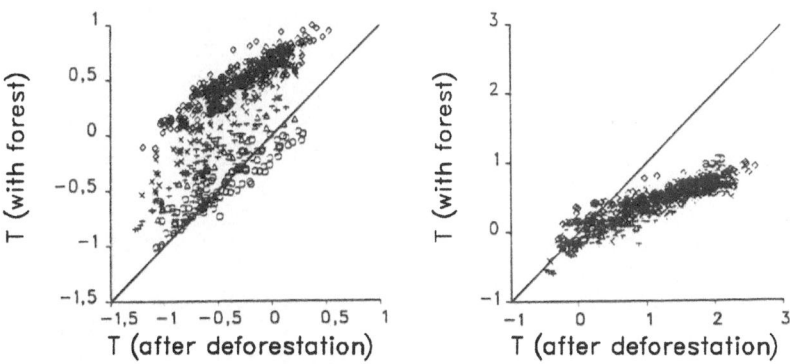

Fig. 4.48. Comparison of temperature for both simulations at ground level (*left*) and at crown height (*right*). Circles $n_c = 0$; triangles $0 < n_c \leqslant 0.2$; plus signs $0.2 < n_c \leqslant 0.4$; crosses $0.4 < n_c \leqslant 0.6$; diamonds $n_c > 0.6$

As a result of the differences in the nocturnal cooling rate, local differences in air temperature, density and pressure are developed. This leads to the establishment of wind systems which are typical for valleys.

Under the influence of gravity the cold air, formed at night over the slopes, moves against the orographic gradients to the lowest point of the region. This results in a nocturnal drainage flow with a vertical thickness of only a few meters

(King 1973). The cold air moving down either side of a valley gathers at the valley floor, resulting in a pressure gradient between the valley and the surrounding plain and this leads to the mountain wind. It is known from observations in different valleys that this phenomenon extends over a larger vertical thickness and may even attain the height of the bordering mountains (Reiter et al. 1984).

These characteristic flow patterns should also be included in the results of the simulations. In Fig. 4.49 the horizontal velocities following the terrain at a level 2 m above ground are shown for both cases by wind vectors. It is clearly evident that the air moves down the slopes, converges along the valley axis and then flows out to the south. In case I the wind speeds are notably reduced by the vegetation. Only in clearings in the forest, where the air cools more than in the surrounding stand, are cold air drainage flows with significant wind speeds developed. After complete deforestation stronger surface winds, with values of

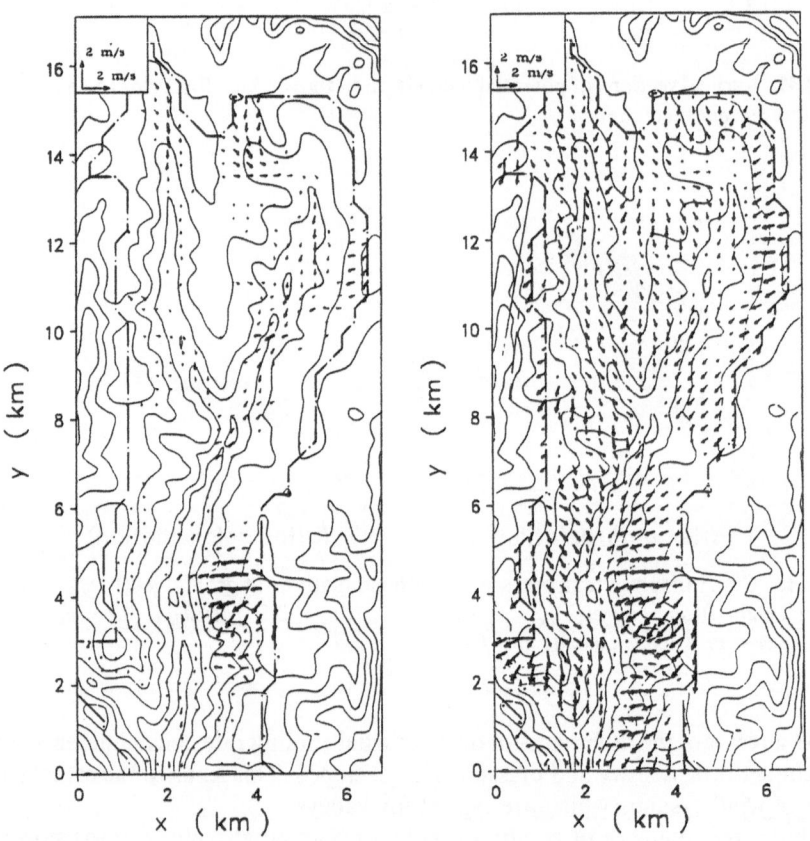

Fig. 4.49. Simulated wind vectors at 2 m above ground prior to deforestation (*left*) and thereafter (*right*)

up to 2.2 m s^{-1} at 2 m above ground, are developed over the entire Finkenbach valley.

Typical vertical profiles of the slope wind for both simulations at point S1 are presented in Fig. 4.50. The wind speed maximum in case I is simulated above the canopy. This coincides approximately with the temperature minimum. Below this the air flow is notably reduced by the stand elements. In the unforested case, the maximum is shifted downward with a pronounced increase. This may be ascribed to the lower temperatures close to the ground and thus to an increase in the temperature deficit.

A vertical integration of the two profiles results in a larger value for case I. Thus, more (but less cooled) air masses flow from the slopes to the valley center than do in the simulation for complete deforestation. The resulting larger horizontal wind convergence along the valley floor leads to higher vertical velocities responsible for the greater cooling of the valley atmosphere, as discussed above. The participation of a thicker layer in the cold air drainage for case I may be explained by the following factors: the strongest cooling takes place in the crown zone, i.e., at a height where turbulence and vertical velocity are already rather pronounced. As a consequence, higher atmospheric levels are cooled more than in case II. At the corresponding slope inclinations these layers contribute to the cold air drainage.

The drainage of cooled air does not take place continuously, but rather in more or less periodic pulsations. Such wind fluctuations have been described by Küttner (1949), Thyson (1968), Manins and Sawford (1979) and Doran and Horst (1981). The periods of these phenomena, sometimes referred to as air avalanches, are in the range 5–75 min.

Figure 4.51 shows the development of the velocity at 2 m above ground over more than 3 h at point S3 after deforestation. The velocity V varies considerably during this period with short-term fluctuations of to 0.7 m s^{-1} within only a few minutes. Numerical effects can be excluded as the time step is between 2 and 3 s and a decrease or increase in this order of magnitude may be followed over 60–100 time steps.

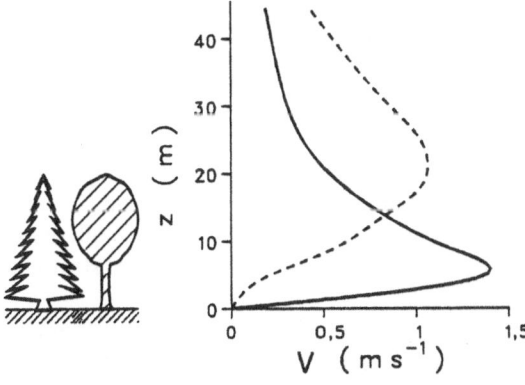

Fig. 4.50. Vertical profiles of downslope wind at site S1 (*dashed line* prior to deforestation; *solid line* thereafter)

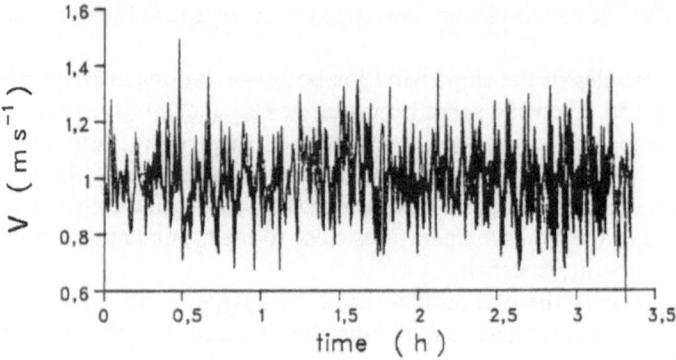

Fig. 4.51. Simulated time record of wind speed at site S3 after deforestation

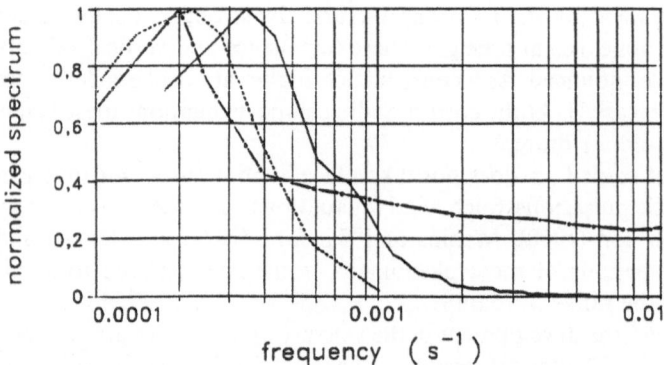

Fig. 4.52. Velocity spectra. *Solid line* Simulated (after deforestation); *dashed* and *dotted line* Doran and Horst (1981) (observed); *dashed line* Tyson (1968) (observed)

The velocity spectrum for this period is presented in Fig. 4.52. For comparison, two spectra obtained from field measurements in other valleys are included. The three profiles show a rather similar shape with a pronounced maximum. While the maxima for the observed velocity fluctuations are 75 and 83 min, the simulation results in a value of 47 min.

The periodicity is controlled to a high degree by the orography of the area and by the thermal stratification (McNider 1982).

In the simulation with forest cover at the same site the phenomenon of cold air pulsation is not developed. The wind speed is generally low and does not show larger variations over the period considered. An explanation is offered by the theory of McNider which does not allow pulsations during neutral stratification, as were simulated inside the forest.

The cold air drainage from the slopes converges at the valley floor and moves south. This mountain wind is much more pronounced than the downslope winds, as a large mass of cold air is removed by it. In Fig. 4.53 the vertical

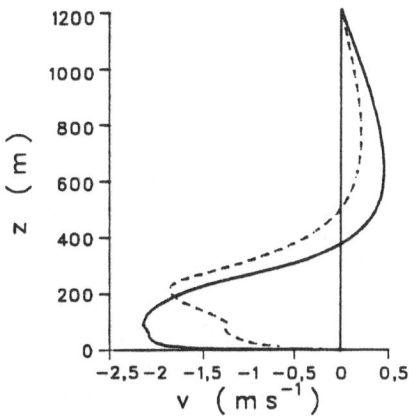

Fig. 4.53. Vertical profiles of mountain wind at site S2 (*dahsed line* prior to deforestation; *solid line* thereafter)

profiles of the mountain wind at the exit of the Finkenbach valley are shown for both cases. For case I a maximum is calculated for a height 200 m above ground with a return flow at a height above the bordering mountains. After deforestation (case II) the mountain wind intensity increases and the maximum is shifted to a lower elevation. Concurrently, the return flow becomes stronger and is developed over a deeper zone.

This leads to the as yet unanswered question whether an unforested or a forested slope contributes more effectively to cold air production. This is of particular importance in regional planning when an optimal fresh air supply to urban areas has to be assured.

Open ground is usually thought to be the main producer of cold air (Weise 1953; Harlfinger 1976). A cold air production rate of $12 \, \text{m}^3 \, \text{m}^{-2} \, \text{h}^{-1}$ was calculated by King (1973) for a shallow unforested slope. This implies that over every square meter of the slope a 12-m-thick layer of cold air is produced per hour. Over forested ground the present rate of $0.6 \, \text{m}^3 \, \text{m}^{-2} \, \text{h}^{-1}$ is assumed by, e.g., DWD.

Field measurements by Hauf and Witte (1985) on the thermal and dynamic structure of nocturnal wind systems in a small tributary valley of the Rhine gave a production rate of $32 \, \text{m}^3 \, \text{m}^{-2} \, \text{h}^{-1}$. This value is almost three times that estimated by King. The field measurements were carried out in an area with about 50% tree cover. These observations suggest that vegetation exerts a pronounced influence on the production of cold air.

The production rate of cold air is determined in field experiments by relating the total mass flux out of the valley to the valley area. This estimation is only possible under low larger-scale pressure gradients so that a passive influence of the orography can be excluded.

The total mass flux M at the southern exit of the Finkenbach valley may be calculated from:

$$M = \int_F \varrho v \, dF. \tag{4.46}$$

The integration is carried out to the height at which the velocity component parallel to the valley's axis (here v) changes the direction. For the two simulations this height is different. For M a value of 8.7×10^5 kg s^{-1} is found in the forested situation while it has a value of 5.6×10^5 kg s^{-1} in the unforested situation. This implies that the forested slopes produce more cold air than the unforested ones. However, the simulation for case II shows that the air close to the ground is much cooler than for case I. In addition, the area of influence of the Finkenbach valley $(8 \times 10^7$ m$^2)$ is required to calculate the cold air production. This area is covered to 78% by forest and only to 22% by short grass.

From these data a cold air production rate of 42.7 m^3 m^{-2} h^{-1} is found for the forest areas and a rate of 25 m^3 m^{-2} h^{-1} for the meadows. The order of magnitude of these simulation results agree well with the observations of Hauf and Witte and conclusively show that the value of 0.6 m^3 m^{-2} h^{-1} for forested slopes is much too low.

An explanation for the observation that forested slopes are more efficient in producing cold air than meadows was given by Gossmann (1984). From satellite pictures he found that the nocturnal surface temperatures of forests and neighboring open ground in the higher parts of the Black Forest differ by 6°C and locally by up to 10°C. This leads to a difference of 30 W m^{-2} in the longwave outgoing radiation. If it is assumed that during the time the observations were being carried out the turbulent latent heat flux was negligible, the energy budget in the canopy region can be reduced to the terms net radiation and turbulent sensible heat flux. As a consequence, the above difference of 30 W m^{-2} in the stand can only be compensated by an increased turbulent heat flux. As the temperature difference between the crown and the air around it is less than between meadows and the atmosphere close to the ground, the cooling rate is also considerably less. Nevertheless, to balance the radiation budget, a larger amount of energy has to be withdrawn from the air which is in contact with the stand elements. This is only possible when a larger volume of air is involved in the cooling process.

In a two-dimensional study on a simple slope the amount by which the vertically integrated mass flux over a stand differs from that over an unforested slope was investigated. For this purpose the development of drainage flow with time was studied for a 4-km-long slope with an inclination of 5°, in one case with trees and in the other case without. All numerical, meteorological and stand-specific input parameters were the same as those used in the preceding three-dimensional simulations. A constant forest density of $n_c = 0.8$ was assumed.

The total downflow at the lower part of the slope was calculated at 15-min intervals throughout the 6-h real-time simulation. For this purpose, the mass flux was integrated from the ground up to the height at which the slope-parallel velocity component changes direction. The calculated ratio of downflow with forest (index W) to the one without forest (index F) is presented in Fig. 4.54. Even on this geometrically simple slope, the volume of air cooled is larger when forest is present than when it is not. Averaged over time, a value of $F_W/F_F = 1.36$ is found. In the three-dimensional simulation, a ratio of 1.56 was

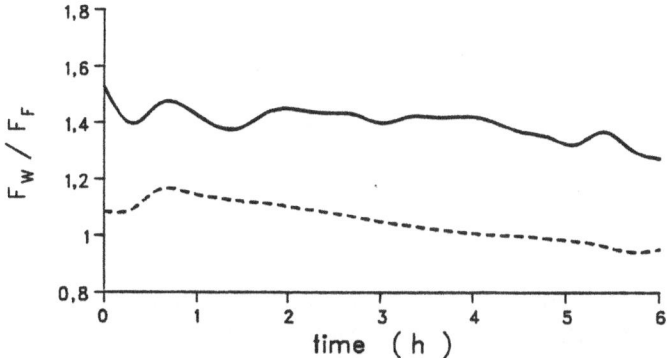

Fig. 4.54. Time evolution of the ratio between forested (F_W) and unforested slopes (F_F) for mass flux (*solid line*) and horizontal heat flux (*dashed line*)

found for the total mass flux from the Finkenbach valley. This larger value may be ascribed to the on-average steeper orography.

A similar evaluation for the ratio between the temperature fluxes or the horizontal heat fluxes does not show a marked preference for a slope with a specific cover. For the forest, the vertically integrated flux is larger, but the temperature disturbance is smaller than in the simulation without trees.

5 Concluding Remarks

The present study is a contribution toward a more complete understanding of the distribution of meteorological variables within the area of influence of stands. The observations presented in the literature are complemented by the results of numerical simulations, thereby allowing an assessment of the complex interactions between plants and air flow.

A comparison between the calculated wind and turbulence patterns and those determined in wind tunnels shows satisfactory agreement. Furthermore, the advantages of a numerical simulation model as compared with laboratory experiments have been outlined. In particular, this results from the extreme variability of the meteorological input parameters and the stand geometry. The simulations also provide consistent three-dimensional distributions of the various variables which can be determined only with great difficulty, if at all, in the laboratory. To verify the models, laboratory and field measurements are obviously required, illustrating the mutual complementarity of the two methods of investigation.

The simulation results for groups and rows of trees facilitate the design of plantations with optimal wind sheltering effects for any site. A consideration of local conditions makes the transfer of experimental data to other sites especially difficult and so, here also, the simulation represents an important contribution to finding the most effective solution.

With the simulation of potential climatic modifications as a result of deforestation or drastic forest decline, on either a local or a regional scale, a new approach was outlined in that, so far, there are virtually no experimental data available on this aspect. This is because for the estimation of potential effects it is necessary to know the situation prior to and after deforestation. Where long-term observations are involved, one would have to cut down the forest in the area under study and repeat the measurements. If it was then found that deforestation had had an adverse effect on the local climate, it would take years to reestablish the original condition and this may not even be possible.

In contrast to this, the stand is easily 'deforested' in the simulation model. From a comparison of the results of a simulation with and without the stand present, qualitative and quantitative conclusions can be drawn as to the possible changes in local climate.

The applicability of this method was demonstrated for the deforestation around the western runway of Frankfurt Airport. Although only two exemplary cases were investigated, and thus there may be certain reservations about the results as far as the climatic changes are concerned, the basic statements agree with the assessment by DWD. As this expertise was supported only partly by field measurements, it is really of a more exploratory nature and agreement with the simulation results is of a more qualitative type. However, the simulation results may also be evaluated in a quantitative manner, thereby aiding decisions necessary to solve related problems.

Another realistic example is presented by changes in cold air drainage patterns and the local climate resulting from hypothetical deforestation in the Finkenbach valley. This led to a new insight into the role of forested slopes for the supply of fresh air to urban areas. While it has previously been assumed that unforested slopes are more efficient sources of cold air production, it was possible to show here that although they do in fact cool the air to a greater degree, the presence of a stand subjects a larger volume of air to the cooling process.

The examples presented here show only a small selection of the potential uses of the numerical simulation model introduced. Especially in the field of urban and regional planning decisions have continuously to be made to optimize utilization of the available stand resources. The present version of the model may be applied directly to wider areas, though it is not suitable for simulations within cities, as for this purpose individual houses and blocks have to be modeled.

For quite a few applications, a detailed knowledge of the vertical distribution of the meteorological variables in the lowermost part of the atmosphere is not essential. The integrated influence of this layer on the·air moving over it is of greater importance. By evaluating the results of simulations like the ones discussed here it should be possible to define parameters for vegetation, including taller trees. This could then be incorporated in other simulation models and in the predictive models routinely used by weather services.

References

Allen LH (1968) Turbulence and wind speed spectra within a Japanese Larch plantation.
J Appl Meteorol 7:73–78

Allison JK, Herrington LP, Morton JD (1968) Wind and turbulent diffusion in and above
a jungle canopy. Bull Am Meteorol Soc 49:765

Amiro BD (1990) Comparison of turbulence statistics within three boreal forest canopies.
Bound Layer Meteorol 51:99–122

Atkinson BW (1981) Meso-scale atmospheric circulations. Academic Press London

Atwater MA, Brown P (1974) Numerical calculation of the latitudinal variation of solar
radiation for an atmosphere of varying opacity. J Appl Meteorol 13:289–297

Baltaxe R (1967) Air flow patterns in the lee of model windbreaks. Arch Met Geophys
Bioklimatol B3:287–312

Baumgartner A (1956) Untersuchungen über den Wärme- und Wasserhaushalt eines
jungen Waldes. Ber Dtsch Wetterdienst Offenbach 28

Baumgartner A (1973) Wald als Umweltfaktor in der Grenzschicht Erde/Atmosphäre.
Ber Meteorol Ges. München 3

Baumgartner A, Kirchner M (1980) Impacts due to deforestation. In: Bach W (ed)
Interactions of energy and climate. Reidel, London, pp 305–316

Beadle CL, Talbot H, Jarvis PG (1982) Canopy structure and leaf area index in a mature
Scots pine forest. Forestry 55:102–123

Becker HG (1978) Eine Hierarchie atmosphärischer Prognosemodelle unter Einschluß
orographisch bedingter Einflüsse. Masterthesis, Meteorol Inst, Univ Mainz

Berg H (1947) Einführung in die Bioklimatologie. Bouvier, Bonn

Bergen JD (1969) Cold air drainage on a forested slope. J Appl Meteorol 8:884–895

Bergen JD (1971) Vertical profiles of windspeed in a pine stand. For Sci 17:314–321

Blackadar AK (1962) The vertical distribution of wind and turbulent exchange in
a neutral atmosphere. J Geophys Res 67:3095–3102

BML (ed) (1984, 1986) Waldschadenserhebung 1984. Bundesministerium für Ernährung,
Landwirtschaft und Forsten (BML)

Braden H (1982) Simulationsmodell für den Wasser-, Energie- und Stoffhaushalt in
Pflanzenbeständen. Ber Inst Meteorol Univ Hannover 23

Bradley EF, Mulhearn PJ (1983) Development of velocity and shear stress distribution in
the wake of a porous shelter fence. J Wind Eng Ind Aerodyn 15:145–156

Brehm M (1986) Experimentelle und numerische Untersuchungen der Hangwind-
schicht und ihrer Rolle bei der Erwärmung von Tälern. Ber Meteorol Inst Univ
München 54

Brooks DL (1950) A tabular method for the computation of temperature change by
infrared radiation in the free atmosphere. J Meteorol 7:313–325

Budagovsky AI, Ross JK, Tooming HG (1968) Actinometry and atmospheric optics.
Valgus, Tallin

Burger H (1951) Waldklimafragen Meteorologische Beobachtungen im Brandiswald. Mitt Schweiz Anst Forst Versuchswes 27:19–75

Businger, JA, Wyngaard JC, Izumi Y, Bradley EF (1971) Relationships in the atmospheric surface layer. J Atmosph Sci 28:181–189

Caughey SJ, Wyngaard JC, Kaimal JC (1979) Turbulence in the evolving stable boundary layer. J Atmosph Sci 36:1041–1052

Chamberlain AC (1975) The movement of particles in plant communities. In: Monteith JL (ed) Vegetation and the atmosphere 1. Academic Press, New York, pp 155–204

Chason JW, Baldocchi DD, Huston MA (1991) A comparison of direct and indirect methods for estimating forest canopy leaf area. Agric For Meteorol 57:107–128

Chroust L (1968) Das Temperaturregime in verschieden durchforsteten Eichen-Stangenhölzern. Allg Forst Jäger Z 139:163–173

Cionco RM (1962) A preliminary model for air flow in the vegetative canopy. Bull Am Meteorol Soc 43:319

Cionco RM (1965) A mathematical model for air flow in a vegetative canopy. J Appl Meteorol 4:517–522

Cionco RM (1971) Application of the ideal canopy flow concept to natural and artificial roughness elements. Tech Rep ECOM-5372, Atm Sci Lab, White Sands Missile Range, NM

Cionco RM (1972) Intensity of turbulence within canopies with simple and complex roughness elements. Bound Layer Meteorol 2:453–465

Cionco RM (1978) Canopy index values for various canopy densities. Bound Layer Meteorol 15:81–93

Cionco RM (1985) Modeling windfields and surface layer wind profiles over complex terrain and within vegetative canopies. In: Hutchison BA, Hicks BB (eds) The forest-atmosphere interaction. Reidel, Dordrecht, pp 501–520

Clarke RH, Dyer AJ, Brooks RR, Reid DG, Troup AJ (1971) The Wangara experiment. Boundary Layer data. Tech Rep 19 Div Meteorol Phys, CSIRO, Melbourne, Aust

Cowan IR (1968) Mass, heat and momentum exchange between stands of plants and their atmospheric environment. QJR Meteorol Soc 94:318–322

CSIRO (1986) Research report 83–85. Tech Rep CSIRO Mordialloc, Vic, Aust

Deardorff JW (1978) Efficient prediction of ground surface temperature and moisture with inclusion of a layer of vegetation. J Geophys Res 83:1889–1903

Defant F (1951) Local winds. Compendium of meteorology. Am Meteorol Soc

Denmead OT (1964) Evaporation sources and apparant diffusivities in a forest canopy. J Appl Meteorol 3:383–389

Denmead OT (1984) Plant physiological methods for studying evapotranspiration: problems of telling the forest from the trees. Agric Water Manag 8:167–189

Denmead OT, Bradley EF (1985) Flux-gradient relationships in a forest canopy. Hutchison BA, Hicks BB (eds) The forest-atmosphere interaction. Reidel, Dordrecht, pp 421–442

Dickerson MH, Gudicksen PH (1983) Atmospheric studies in complex terrain. Tech Rep ASCOT 84-1, LLNL, Livermore, Cal

Dolman AJ, van den Burg GJ (1988) Stomatal behaviour in an oak canopy. Agric For Meteorol 43:99–108

Doran JC, Horst TW (1981) Velocity and temperature oscillations in drainage winds. J Appl Meteorol 20:361–364

Dutton JA (1976) The ceaseless wind. McGraw-Hill, New York

DWD Deutscher Wetterdienst (ed) (1965) Gutachten über mögliche klimatische Auswirkungen eines beim Flughafen Frankfurt vorgesehenen Waldeinschlages. Deutscher Wetterdienst Offenbach

DWD Deutscher Wetterdienst (ed) (1966) Gutachten über die mögliche klimatische Auswirkung einer Verminderung des Waldgürtels am Nordrand des Frankfurter Flughafens. Deutscher Wetterdienst Offenbach

DWD Deutscher Wetterdienst (ed) (1967 Zusatzgutachten zu hiesigem Gutachten vom Oktober 1965 über die möglichen klimatischen Auswirkung eines beim Flughafen Frankfurt a. M. vorgesehenen Waldeinschlags. Deutscher Wetterdienst Offenbach

Egger J (1987) Simple models of the valley-plain circulation Part I. Minimum resolution model. Meteorol Atmosph Phys 36:231–242

Eimern J van (1957) Über die Veränderlichkeit der Windschutzwirkung einer Doppelbaumreihe bei verschiedenen meteorologischen Bedingungen. Ber Deutscher Wetterdienst Offenbach 28

Eimern J van, Karschon R, Razumova LA, Robertson GW (1964) Windbreaks and shelterbelts. Tech Rep 59 WMO

Finnigan JJ, Bradley EF (1983) The turbulent kinetic energy budget behind a porous barrier: an analysis in streamline coordinates. J Wind Eng Ind Aerodyn 15:157–168

Finnigan JJ, Mulhearn PJ (1978) A simple mathematical model of airflow in waving plant canopies. Bound Layer Meteorol 14:415–431

Flemming G (1964) Das Klima an Waldbestandsrändern. Ber 71 Hydrologischer Dienst DDR, Berlin

Flüggen C (1991) Die Evapotranspiration von Kiefern unter Berücksichtigung des Grundwasserflurabstandes. Ber Inst Meteorol Klimatol Univ Hannover 39

Fons WL (1940) Influence of forest cover on wind velocity. J For 38:481–486

Freytag C (1985) MERKUR-results: aspects of the temperature field and the energy budget in a large Alpine valley during mountain and valley wind. Beitr Phys Atmosph 58:458–476

Freytag C (1987) Results from the MERKUR-experiment: mass budget and vertical motions in a large valley during mountain and valley wind. Meteorol Atmosph Phys. 37:129–140

Friedlander SK, Johnstone HF (1957) Deposition of suspended particles from turbulent gas streams. Ind Eng Chem 49:1151–1156

Fritschen LJ (1985) Characterization of boundary conditions affecting forest environmental phenomena. Hutchison BA, Hicks BB (eds) The forest-atmosphere-interaction. Reidel, Dordrecht, pp 3–23

Fritschen LJ, Driver CH, Avery C, Buffo J, Edmonts R, Kinerson R, Schiess P (1969) Dispersion of air tracers into and within a forested domain. Tech Rep Atmosph Sci Lab Ft Huachuca, AZ

Gal-Chen T, Sommerville RC (1975) Numerical solution of the Navier–Stokes equations with topography. J Comp Phys 17:276–309

Garratt AJ (1983) Drainage flow prediction with a one-dimensional model including canopy, soil and radiation parameterization. J Appl Meteorol 22:79–91

Geiger R (1926) Untersuchungen über das Bestandsklima. Forstwiss Centralbl 48:337–349

Geiger R (1950) Das Klima der bodennahen Luftschicht, 3rd edn. Vieweg, Braunschweig

Geiger R (1961) Das Klima der bodennahen Luftschicht, 4th edn. Vieweg, Braunschweig

Geiger R, Amann H (1931) Forstmeteorologische Messungen in einem Eichenbestand. Forstwiss Centralbl 53:341–351

Gill AE (1982) Atmosphere-ocean dynamics. Academic Press, New York

Glaab H (1986) Lagrangsche Simulation der Ausbreitung passiver Luftbeimengungen in inhomogener atmosphärischer Turbulenz. Meteorol Inst TH Darmstadt

Göhre K, Lützke R (1956) Der Einfluß von Bestandsdichte und -struktur auf das Kleinklima im Walde. Arch Forstwes 5:487–572

Gossmann H (1984) Satelliten-Thermalbilder Ein neues Hilfsmittel für die Umweltforschung. Ber 16 Bundesanst Landeskd Raumordnung, Bonn

Grant HG (1983) The scaling of flow in vegetative structures. Bound Layer Meteorol 27:171–184

Grin AM, Rauner JL, Utekhin VD (1970) Proc USSR Acad Sci Geophys Ser 4:10–23

Gross G (1984) Eine Erklärung des Phänomens Maloja-Schlange mittels numerischer Simulation. PhD Thesis, Meteorol Inst TH Darmstadt

Gross G (1985) Numerische Simulation nächtlicher Kaltluftabflüsse und Tiefsttemperaturen in einem Moselseitental. Meteorol Rundsch 38:161–171

Gross G (1986) A numerical study of the land and sea breeze including cloud formation. Beitr Phys Atmosph 59:97–114

Gross G (1987) Effects of deforestation on local climate and nocturnal drainage flow. Bound Layer Meteorol 37:315–339

Gross G (1988) A numerical estimation of the deforestation effects on local climate in the area of the Frankfurt International Airport. Beitr Phys Atmosph 61:219–231

Gross G (1989) Numerical simulation of the nocturnal flow systems in the Freiburg area for different topographies. Beitr Phys Atmosph 62:57–72

Gross G (1991) Anwendungsmöglichkeiten mesoskaliger Simulationsmodelle dargestellt am Beispiel Darmstadt, Teil 1: Wind- und Temperaturfelder. Meteorol Rundsch 43:97–112

Gross G. Wippermann F (1987) Channeling and counter-current in the Upper-Rhine valley. J Appl Meteorol 26:1293–1304

Gross G, Vogel H, Wippermann F (1987) Dispersion over and around a steep obstacle for varying thermal stratification-numerical simulations. Atmosph Environ 21:483–490

Hack JJ, Schubert WH (1981) Lateral boundary conditions for tropical cyclone models. Mon Weather Rev 109:1404–1420

Halldin S (1985) Leaf and bark area distribution in a pine forest. In: Hutchison BA, Hicks BB (eds) The forest-atmosphere interaction. Reidel, Dordrecht, pp 39–58

Hanna SR (1981) Applications in air pollution modelling. In: Nieuwstadt FTM, Van Dop H (eds) Atmospheric turbulence and air pollution modelling. Reidel, Dordrecht, pp 275–310

Harlfinger O (1976) Die bioklimatische Bedeutung des 'Höllentälers' für Freiburg. Meteorol Rundsch 26:15–18

Hauf T, Witte N (1985) Fallstudie eines nächtlichen Windsystems. Meteorol Rundsch 38:33–42

Heisler GM (1985) Measurements of solar radiation on vertical surfaces in the shade of individual trees. In: Hutchison BA, Hicks BB (eds) The forest-atmosphere interaction. Reidel, Dordrecht, pp 319–335

Hennemuth B (1987) Heating of a small alpine valley. Meteorol Atmosph Phys 36:287–296

Hennemuth B, Schmidt H (1985) Wind phenomena in the Dischma valley during DISKUS. Arch Meteorol Geophys. Bioklimatol B35:361–387

Herbst W (1965) Filter- und Schutzwirkung des Waldes gegen radioaktive und andere Beimengungen der Atmosphäre, Allg Forstz 20:216–220

Hicks BB, Hyson P, Moore CJ (1975) A study of eddy fluxes over a forest. J Appl Meteorol 14:58–66

Hicks BB, Hess GD, Wesely ML (1979) Analysis of flux-profile relationships above tall vegetation – an alternative view. QJR Meteorol Soc 105:1074–1077

Hosker RP, Nappo CJ, Hanna SR (1974) Diurnal variation of vertical thermal structure in a pine plantation. Agric Meteorol 13:259–265

Hoyningen-Huene J (1980) Mikrometeorologische Untersuchungen zur Evapotranspiration von bewässerten Pflanzenbeständen. Ber Inst Meteorol Univ Hannover 19

Impens I, Lemeur R (1969) Extinction of net radiation in different crop canopies. Arch Meteorol Geophys Bioklimatol B17:403–412

Inoue K (1963) On the turbulent structure of airflow within crop canopies. J Meteorol Soc Jpn 11:18–22

Jackson PS (1981) On the displacement height in the logarithmic velocity profile. J Fluid Mech 111:15–26

Jarvis PG, James GB, Landsberg JJ (1975) Coniferous forest. In: Monteith JL (ed) Vegetation and the atmosphere 2. Academic Press, New York, 171–240

Keller T (1970) Zum Problem der verkehrsbedingten Bleirückstände in der Vegetation. Straße Verkehr 1:117–121

King E (1973) Untersuchungen über kleinräumige Änderungen des Kaltluftabflusses und der Frostgefährdung durch Straßenbauten. Ber 130 Deutscher Wetterdienst Offenbach

Kittridge J (1948) Forest influences. McGraw Hill, New York

Klemp JB, Lilly DK (1975) The dynamics of wave induced downslope winds. J Atmosph Sci 32:320–329

Klemp JB, Lilly DK (1978) Numerical simulation of hydrostatic mountain waves. J Atmosph Sci 35:78–107

Koch HG (1961) Die warme Hangzone. Z Meteorol 15:151–171

Koch W (1987) The influence of the forest decay on local climate variables. Tech Rep Meteorol Inst TH Darmstadt

Kondo J, Akashi A (1976) Numerical studies of the two-dimensional flow in horizontally homogeneous canopy layers. Bound Layer Meteorol 10:255–272

Kontratyev K (1969) Radiation in the atmosphere. Academic Press, New York

Kuhn PM (1963) Radiometersonde observations of infrared flux emissivity of water vapour. J Appl Meteorol 2:368–378

Kurata K (1982) Theoretische Untersuchungen der Turbulenz innerhalb eines Pflanzenbestandes. Ber Inst Meteorol Univ Hannover 20

Kurz H (1977) Turbulente Diffusion in einer atmosphärischen Grenzschicht mit Rossby-Zahl-Ähnlichkeit. PhD Thesis, Meteorol Inst TH Darmstadt

Küttner J (1949) Periodische Luftlawinen. Meteorol Rundsch 2:183–184

Ladefoged K (1963) Transpiration of forest trees in closed stands. Physiol Plant 16:378–414

Landsberg JJ, Thom AS (1971) Aerodynamic properties of a plant of complex structures. QJR Meteorol Soc 97:565–570

Landsberg JJ, Jarvis PG, Slater MB (1973) The radiation regime in a spruce forest. In: Slateyer RO (ed) Plant response to climatic factors. UNESCO, Paris, pp 411–418

Lang ARG, Yueqin X (1986) Estimation of leaf area index from transmission of direct sunlight in discontinuous canopies. Agric For Meteorol 37:229–243

Larcher W (1984) Ökologie der Pflanzen. Ulmer, Stuttgart

Legg BJ (1983) Turbulent dispersion from an elevated line source: Markov-chain simulations of concentrations and flux-profiles. QJR Meteorol Soc 109:645–600

Legg BJ, Raupach MR (1982) Markov-chain simulation of particle dispersion in inhomogeneous flows: the mean drift velocity induced by a gradient in Eulerian velocity variance. Bound Layer Meteorol 24:3–13

Lehn WH (1991) A two-band clear sky albedo model for a pine forest. Meteorol Rundsch 43:129–139

Lewellen WS, Teske ME, Sheng YP (1979) Micrometeorological applications of a second order closure model of turbulent transport. In: 2nd Symp Turbulent shear flows, Imperial College, London

Lindroth A, Halldin S (1990) Gradient measurements with fixed and reversing temperature and humidity sensors above a thin forest. Agric For Meteorol 53:81–104

Lützke E (1961) Das Temperaturklima von Waldbeständen und -lichtungen im Vergleich zur offenen Feldflur. Arch Forstwes 18:921–927

Lützke E (1967) Die Luftfeuchtigkeit im Walde im Vergleich zur offenen Feldflur. Arch Forstwes 24:629–633

Mahrer Y, Pielke RA (1977a) The effects of topography on sea and land breezes in a two-dimensional numerical model. Mon Weather Rev 105:1151–1162

Mahrer Y, Pielke RA (1977b) A numerical study of the airflow over irregular terrain. Beitr Phys Atmosph 50:98–113

Manins PC, Sawford BL (1979) Katabatic winds: a field case study. QJR Meteorol Soc 105:1011–1025

Martin P (1989) The significance of radiative coupling between vegetation and the atmosphere. Agric For Meteorol 49:45–54

Mason PJ, Sykes RI (1979) A simple Cartesian model of boundary layer flow over topography. J Comp Phys 28:198–210

Massman WJ, Kaufmann MR (1991) Stomatal response to certain environmental factors: a comparison of models for subalpine trees in the Rocky Mountains. Agric For Meteorol 54:155–167

Mayer H (1982) Forstmeteorologie. Promet, Deutscher Wetterdienst Offenbach 3

Mayer H (1985) Baumschwingungen und Sturmgefährdung des Waldes. Tech Rep 51 Meteorol Inst Univ München

Mayhead GJ (1973) Some drag coefficients for British forest trees derived from wind tunnel studies. Agric Meteorol 12:123–130

McAneney KJ, Barber RF, Salinger MJ, Porteous AS (1989) Modification of an orchard climate with increasing shelterbelt height. Agric For Meteorol 49:177–190

McBean GA (1968) An investigation of turbulence within the forest. J Appl Meteorol 7:410–416

McCaughey JH (1985) Energy balance storage terms in a mature mixed forest at Petawana, Ontario—a case study. Bound Layer Meteorol 31:89–101

McCumber MC (1980) A numerical simulation of the influence of heat and moisture fluxes upon mesoscale circulations. PhD Thesis, Univ Virginia Charlottesville

McDonald J (1960) Direct absorption of solar radiation by atmospheric water vapor. J Meteorol 7:319–328

McNider RT (1982) A note on velocity fluctuations in drainage flows. J Atmosph Sci 39:1658–1660

Mellor GL, Yamada T (1982) Development of a turbulence closure model for geophysical fluid problems. Rev Geophys Space Phys 20:851–875

Meroney RN (1968) Characteristics of wind and turbulence in and above model forests. J Appl Meteorol 7:780–788

Meyers TP, Baldocchi DD (1990) The budgets of turbulent kinetic energy and Reynolds stress within and above a deciduous forest. Agric For Meteorol 53:207–222

Miller DR, Lin JD (1985) Canopy architecture of a red maple edge stand measured by a point drop method. In: Hutchison BA, Hicks BB (eds) The forest-atmosphere interaction. Reidel, Dordrecht, pp 59–70

Miller MJ, Thorpe AJ (1981) Radiation conditions for the lateral boundaries of limited-area numerical models. QJR Meteorol Soc 107:615–628

Miller DR, Lin JD, Lu ZN (1991) Air flow across an Alpine forest clearing: a model and field measurements. Agric For Meteorol 56:209–226

Mitscherlich G (1970) Wald, Wachstum und Umwelt Band 1. Form und Wachstum von Baum und Bestand. Sauerländer, Frankfurt

Mitscherlich G (1971) Wald, Wachstum und Umwelt Band 2. Waldklima und Wasserhaushalt. Sauerländer, Frankfurt

Mitscherlich G (1973) Wald und Wind. Allg Forst Jäger Z 144:76–81

Mitscherlich G (1975) Wald, Wachstum und Umwelt Band 3. Boden, Luft Produktion. Sauerländer, Frankfurt

Mitscherlich G, Künstle E, Lang W (1967) Ein Beitrag zur Frage der Beleuchtungsstärke im Bestande. Allg Forst Jäger Z 138:213–223

Monteith JL (1965) Evaporation and environment. Symp Soc Exp Biol 19:205–234

Monteith JL (1973) Principles of environmental physics. Arnold, London

Munro DS (1985) Internal consistency of the Bowen ratio approach to flux estimation over forested wetland. In: Hutchison BA, Hicks BB (eds) The forest-atmosphere interaction. Reidel, Dordrecht, pp 395–404

Nägeli W (1946) Weitere Untersuchungen über die Windverhältnisse im Bereich von Windschutzanlagen. Mitt Schweiz Anst Forstl Versuchswes 24:659–737

Nägeli W (1953) Untersuchungen über die Windverhältnisse im Bereich von Schilfrohrwänden. Mitt Schweiz Anst Forstl Versuchswes 29:213–266

Nägeli W (1954) Die Windbremsung durch einen großen Waldkomplex. IUFRO

Nägeli W (1965) Über die Windverhältnisse im Bereich gestaffelter Windschutzstreifen. Mitt Schweiz Anst Forstl Versuchswes 41:220–300

Neuberger H, Hosler CL, Kockmond WC (1967) Vegetation as aerosol filters. In: Tromp SW, Weihe WH (eds) Biometeorology. Pergamon, Oxford, pp 693–702

Nizinski J, Saugier B (1989) A model of transpiration and soil-water balance for a mature oak forest. Agric For Meteorol 47:1–17

Nord D (1991) Shelter effects of vegetation belts-results of field measurements. Agric For Meteorol 54:363–386

Novak MD (1990) An approximate analytical theory to predict the diurnal effects of local advection. Agric For Meteorol 51:159–176

Oliger J, Sandström A (1976) Theoretical and practical aspects of some initial boundary value problems in fluid dynamics. Tech Rep STAN-CS-76-578 Comp Sci Dep, Stanford Univ

Oliver HR (1971) Wind profiles in and above a forest canopy. QJR Meteorol Soc 97:548–553

Orlanski I (1976) A simple boundary condition for unbounded hyperbolic flows. J Comp Phys 21:251–269

Panggabean H (1978) Schwingungsverhalten von turmartigen Tragwerken unter aero-dynamischer Belastung. Beitr Anwend Aerolast Bauwes 10: 23–45

Papke HE, Krahl-Urban B, Peters K, Schimansky C (1986) Waldschäden. Tech Rep Projektträgerschaft für Biologie, Ökologie und Energie der KFA Jülich

Patrinos AAN, Kistler AL (1977) A numerical study of the Chicago lake breeze. Bound Layer Meteorol 12: 93–123

Paw UKT (1992) A discussion of the Penman form equations and comparisons of some equations to eliminate latent energy flux density. Agric For Meteorol 57: 297–304

Penman HL, Long IF (1960) Weather in wheat: an essay in micrometeorology. QJR Meteorol Soc 86: 16–50

Pielke RA (1984) Mesoscale meteorological modeling. Academic Press, Orlando

Plate E, Quraishi AA (1965) Modeling of velocity distribution inside and above tall crops. J Appl Meteorol 4: 400–408

Prandtl L (1965) Führer durch die Strömungslehre. Vieweg, Braunschweig

Rauner JL (1976) Deciduous Forests. In: Monteith JL (ed) Vegetation and the atmosphere 2. Academic Press, New York, pp 241–264

Raupach MR, Legg BJ (1984) The uses and limitations of flux-gradient relationships in micrometeorology. Agric Water Manag 8: 119–131

Raupach MR, Thom AS (1981) Turbulence in and above plant canopies. Annu Rev Fluid Mech 13: 97–129

Raynor GS (1971) Wind and temperature structure in a coniferous and a contiguous field. For Sci 17: 351–363

Raynor GS, Ogden EC, Hayes JV (1970) Dispersion and deposition of ragweed pollen from experimental sources. J Appl Meteorol 6: 885–895

Raynor GS, Hayes JV, Ogden EC (1974) Particulate dispersion into and within a forest. Bound Layer Meteorol 7: 429–456

Reifsnyder W (1950) Wind profiles in a small isolated forest stand. For Sci 1: 289–297

Reiter R, Müller H, Sladkovic R, Munzert K (1984) Aerologische Untersuchungen der tagesesperiodischen Gebirgswinde unter besonderer Berücksichtigung des Windfeldes im Talquerschnitt. Meteorol Rundsch 37: 176–190

Roache PJ (1972) Computational fluid dynamics. Hermosa, Albuquerque

Rodi W (1980) Turbulence models and their applications in hydraulics. Tech Rep Div Exp Math Fluid Dyn, Delft

Rouhiainen PO, Stachiewicz JW (1979) On the deposition of small particles from turbulent streams. J Heat Transf 101: 166–177

Ruck B, Schmitt F (1986a) Das Strömungsfeld der Einzelbaumumströmung. Forstwiss Centralbl 105: 178–196

Ruck B, Schmitt F (1986b) Umströmung von Koniferen und deren Einfluß auf die Schadstoffdeposition durch Feinsttröpfchen. In: 2. Statuskoll des PEF, Kernforschungszentrum, Karlsruhe, pp 689–702

Ruck R, Adams E (1991) Fluid mechanical aspects of the pollutant transport to coniferous trees. Bound Layer Meteorol 56: 163–196

Rutter AJ (1975) The hydrological cycle in vegetation. In: Monteith JL (ed) Vegetation and the atmosphere 1. Academic Press, New York, pp 111–154

Sadeh WZJE, Cermak E, Kawatani T (1971) Flow over high roughness elements. Bound Layer Meteorol 1: 321–344

Sceicz G (1975) Instruments and their exposure. In: Monteith JL (ed) Vegetation and the atmosphere 1. Academic Press, New York, pp 229–274

Schlichting H (1958) Grenzschichttheorie. Braun, Karlsruhe

Schnelle F (1963) Frostschutz im Pflanzenbau. BLV, München

Schnelle F (1972) Lokalklimatische Studien im Odenwald. Ber 128 Deutscher Wetterdienst Offenbach

Schöpfer W, Hradetzky J (1984) Der Indizienbeweis: Luftverschmutzung maßgebliche Ursache für Walderkrankung. Forstwiss Centralbl 103:231–248

Schuepp PH (1982) Laboratory studies on dry deposition of submicron-site particles on coniferous foliage. Bound Layer Meteorol 24:465–480

Schumann U, Volkert H (1984) Three-dimensional mass- and momentum-consistent Helmholtz equation in terrain-following coordinates. In: Notes on numerical fluid mechanics. GAMM Worksh, Kiel, Vieweg, Braunschweig

Seginer I, Mulhearn PJ, Bradley EF, Finnigan JJ (1976) Turbulent flow in a model plant canopy. Bound Layer Meteorol 10:423–453

Shaw RH, Silverside RH, Thurtell GW (1974) Some observations of turbulence and turbulent transport within and above forest canopies. Bound Layer Meteorol 5:429–449

Shaw RH, Paw UKT, Zhang XJ, Gao W, den Hartog G, Neumann HH (1990) Retrieval of turbulent pressure fluctuations at the ground surface beneath a forest. Bound Layer Meteorol 50:319–338

Stewart JB, Thom AS (1973) Energy budgets in pine forest. QJR Meteorol Soc 99:154–170

Tangermann-Dlugi G (1982) Numerische Simulation atmosphärischer Grenzschichtströmungen über langgestreckten mesoskaligen Hügelketten bei neutraler thermischer Schichtung. Ber 2 Meteorol Inst Univ Karlsruhe

Terpitz W (1981) Ergänzende klimatologische Stellungnahme zur Erweiterung des Flughafens Frankfurt a.M. (Startbahn West). Tech Rep Deutscher Wetterdienst Offenbach

Thom AS (1968) The exchange of momentum, mass and heat between an artificial leaf and the airflow in a wind-tunnel. QJR Meteorol Soc 94:44–55

Thom AS (1971) Momentum absorption by vegetation. QJR Meteorol Soc 97:414–428

Thom AS (1975) Momentum, mass and heat exchange of plant communities. In: Monteith JL (ed) Vegetation and the atmosphere 1. Academic Press, New York, pp 57–110

Trela M (1982) Deposition of droplets from turbulent streams. Wärme Stoffübertragung 16:161–168

Tyson PD (1968) Velocity fluctuations in the mountain wind. J Atmosph Sci 25:381–384

Uchijiama Z (1961) On characteristics of heat balance of water layer under paddy plant cover. Bull Natl Inst Agrie Sci 8:243–263

Uchijima Z (1962) Studies on the micro-climate within plant communities. 1. On the turbulent transfer coefficient within plant layers. J Agric Meteorol 18:1–9

Ulrich W (1987) Simulation von thermisch induzierten Winden und Überströmungssituationen. Ber 57 Meteorol Inst Univ München

Vogel H (1986) Berechnung von Konzentrationsverteilungen mit einem Lagrange Model für zweidimensionale Strömungsfelder. Masterthesis, Meteorol Inst TH Darmstadt

Walmsley JL, Salmon JR, Taylor PA (1982) On the application of a model of boundary-layer flow over low hills to real terrain. Bound Layer Meteorol 23:17–46

Wang YP, Jarvis PG (1988) Mean leaf angles for the ellipsoidal inclination angle distribution. Agric For Meteorol 43:319–321

Wang YP, Jarvis PG (1990) Description and validation of an array model–MAESTRO. Agric For Meteorol 51:257–280

Weise R (1953) Kaltluftstraßen im Weinberg und ihre Auswirkungen. Dtsch Weinbau 8:348

Whittaker H, Likens GE (1975) The biosphere and man. In: Lieth H, Whittaker R (eds) Primary productivity of the biosphere. Springer, Berlin Heidelberg New York, pp 305–328

Wilson JD (1985) Numerical studies on flow through wind-break. J Wind Eng Ind Aerodyn 21:119–154

Wilson JD (1987) On the choice of a windbreak porosity profile. Bound Layer Meteorol 38:37–49

Wilson NR, Shaw RH (1977) A higher order closure model for canopy flow. J Appl Meteorol 16:1197–1205

Wippermann F (1973) Numerical study on the effects controlling the low-level jet. Beitr Phys Atmosph 46:137–154

Wippermann F (1984) Air flow over and in broad valleys: channeling and counter-current. Beitr Phys Atmosph 57:92–105

Wippermann F, Gross G (1981) On the contruction of orographically influenced wind roses for given distributions of the larger-scale wind. Beitr Phys Atmosph 54:492–501

Woelfle M (1937) Verhagerungserscheinungen. Forstwiss Centralbl 81:757–769

Woelfle M (1950) Waldbau und Forstmeteorologie. Bayr Landwirtschaftsverlag, München

Woodruff MP (1956) The spacing interval for supplement shelter belts. J For 54:115–122

Wooldridge G, Thorson PA, Furman RW (1982) Airflow patterns and momentum flux profiles around an isolated mountain. In: DWD (ed) Annalen der Meteorologie. Deutscher Wetterdienst Offenbach, pp 140–142

Yamada T (1982) A numerical model study of turbulent air flow in and above a forest canopy. J Meteorol Soc Jpn 60:439–454

Yamada T (1985) Numerical simulation of the Night 2 data of the 1980 ASCOT experiment in the California Geysers area. Arch Met Geophys Bioklimatol A34:223–247

Yerg DG (1990) Low frequency and fluctuations within an irregular forest. Agric For Meteorol 51:123–144

Ylinen A (1952) Über die mechanische Schaftform der Bäume. Silva Fennica 76

Zhenjia L, Lin JD, Miller DR (1990) Air flow over and through a forest edge: a steady-state numerical simulation. Bound Layer Meteorol 51:179–198

Zimmermann LI (1969) Atmospheric wake phenomena near the canarian Islands. J Appl Meteorol 8:896–907

Symbols

a	leaf area
a_g	albedo of the ground
a_t	tree albedo
A_h	slope azimuth
A_s	solar azimuth
b	leaf area density
Bo	Bowen ratio
c	concentration
c_d	drag coefficient
c_0	ground concentration
c_p	specific heat at constant pressure
d	zero displacement
D	crown diameter
e	vapor pressure
E	turbulent kinetic energy
f, f^*	Coriolis parameters
F	total drag force
g	acceleration due to gravity
h	orography
h_s	trunk height
h_t	tree height
H	top of the model
i_u	turbulence intensity
k	von Karman's constant
k_c	extinction coefficient
K	crown height
K_m, K_h	turbulent diffusion coefficient for momentum and heat, respectively
l	mixing length
l_∞	asymptotic mixing length
L	leaf area index
L_s	latent heat of vaporization
L_*	Monin–Obukhov length scale
M	mass flux
n_c	fraction of the area covered with trees
p	pressure

p_h	hydrostatic part of pressure
P	porosity
P_s	transpiration
Pr_t	turbulent Prandtl number
Q_B	soil heat flux
Q_D	horizontal heat flux
Q_H	sensible heat flux
Q_M	metabolic energy transformation
Q_P	energy flux into biochemical storage
Q_V	latent heat flux
r	optical path length for water vapor
r_g	boundary layer diffusion resistance (leaf)
r_k	cuticular diffusion resistance (leaf)
r_s	stomatal diffusion resistance (leaf)
Re	Reynolds number
R_L	longwave radiation flux
R_{Li}	Lagrangian autocorrelation function
R_N	net radiation flux
R_{NP}	net radiation flux in a canopy
s	specific humidity
s^*	saturation humidity
s_*	water vapor scale
S	direct solar radiation
S_0	solar constant
t	time
t_h	solar hour
T	temperature
T_{Li}	Lagrangian time scale
T_s	temperature in the soil
u, v, w	velocity components in x, y, z direction, respectively
u_i	velocity components in the ith direction ($i = 1, 2, 3$)
u_g, v_g	components of the geostrophic wind
u_*	friction velocity
V	wind speed
w^*	vertical velocity in the (x, y, z^*) coordinate system
W_{FC}	field capacity
W_G	water content of the soil
x, y, z	Cartesian coordinates
x_i	coordinates in the ith direction ($i = 1, 2, 3$)
z_0	roughness length
z_s	depth in the soil
z^*	vertical coordinate in the terrain following coordinate system
Z	zenith angle
α	slope angle
α_t	canopy index

α^*	capillarity of the soil
Γ	psychrometer constant
δ	solar declination
$\Delta x, \Delta y, \Delta z$	grid length in x, y, z direction
Δt	time increment
ε	emissivity
ε_t	emissivity of the stand
θ	potential temperature
θ_0	potential temperature at the ground
θ_*	temperature scale
λ_s	heat conductivity of the soil
ν_u	kinematic viscosity
ν_T	molecular diffusivity of temperature
ϱ	air density
ϱ_w	water density
σ	Stephan-Boltzmann constant
σ_{ui}	velocity variances
τ	shear stress
φ	geographic latitude
Φ	profile function
Ω	random number

Subject Index